Structural Integrity

Volume 14

The *Structural Integrity* book series is a high level academic and professional series publishing research on all areas of Structural Integrity. It promotes and expedites the dissemination of new research results and tutorial views in the structural integrity field.

The Series publishes research monographs, professional books, handbooks, edited volumes and textbooks with worldwide distribution to engineers, researchers, educators, professionals and libraries.

Topics of interested include but are not limited to:

- Structural integrity
- Structural durability
- Degradation and conservation of materials and structures
- Dynamic and seismic structural analysis
- Fatigue and fracture of materials and structures
- Risk analysis and safety of materials and structural mechanics
- Fracture Mechanics
- Damage mechanics
- Analytical and numerical simulation of materials and structures
- Computational mechanics
- Structural design methodology
- Experimental methods applied to structural integrity
- Multiaxial fatigue and complex loading effects of materials and structures
- Fatigue corrosion analysis
- Scale effects in the fatigue analysis of materials and structures
- Fatigue structural integrity
- Structural integrity in railway and highway systems
- Sustainable structural design
- Structural loads characterization
- Structural health monitoring
- Adhesives connections integrity
- Rock and soil structural integrity.

**** Indexing: The books of this series are submitted to Web of Science, Google Scholar and Springerlink ****

This series is managed by team members of the ESIS/TC12 technical committee.

Springer and the Series Editors welcome book ideas from authors. Potential authors who wish to submit a book proposal should contact Dr. Mayra Castro, Senior Editor, Springer (Heidelberg), e-mail: mayra.castro@springer.com

More information about this series at http://www.springer.com/series/15775

Sadjad Pirmohammad · Majid Reza Ayatollahi

Fracture Behavior of Asphalt Materials

 Springer

Sadjad Pirmohammad
Department of Mechanical Engineering
Faculty of Engineering
University of Mohaghegh Ardabili
Ardabil, Iran

Majid Reza Ayatollahi
Fatigue and Fracture Research Laboratory
School of Mechanical Engineering
Iran University of Science and Technology
Tehran, Iran

ISSN 2522-560X ISSN 2522-5618 (electronic)
Structural Integrity
ISBN 978-3-030-39976-4 ISBN 978-3-030-39974-0 (eBook)
https://doi.org/10.1007/978-3-030-39974-0

This Springer imprint is published by the registered company Springer Nature Switzerland AG
The registered company address is: Gewerbestrasse 11, 6330 Cham, Switzerland

Preface

Fracture mechanics is a branch of solid mechanics which studies the propagation of cracks in materials. Indeed, it evaluates the resistance of cracked bodies to fracture using different analytical, numerical, and experimental methods. Fracture mechanics is an important approach to improve the performance of cracked parts. It employs the physics of stress and strain behavior of materials, particularly the theories of elasticity and plasticity, to the microscopic crystallographic defects found in real materials for predicting the macroscopic mechanical behavior of those parts. There is no analytical solution for most of the crack problems with complex geometry and loading conditions. Hence, the analytical solution is only restricted to simple engineering problems. On the other hand, experimental methods are sometimes costly, time-consuming, and laborious. However, numerical methods like finite element, discrete element, and extended finite element are reasonable ways to solve the problems with complicated geometry, loading, and boundary conditions.

Asphalt pavements are layered systems with high quality and expensive materials used in the top layer. There are various distresses such as rutting, corrugation, shoving, cracking, etc. which influence the performance of asphalt pavements. Among these distresses, cracking is a major threat to the performance of asphalt pavements and imposes significant costs on the road agencies. Good understanding of the cracking mechanisms can be an appropriate way to reduce the costs of maintenance processes. Fracture mechanics is a useful and efficient methodology for considering the crack growth behavior of asphalt mixtures.

This book consists of six chapters. An introduction to asphalt concretes together with introducing different types of cracks appearing in asphalt pavements are given in Chap. 1. Chapter 2 discusses the computational modeling of crack growth behavior of asphalt concretes using different methodologies such as discrete element method (DEM), finite element method (FEM), and extended finite element method (XFEM). Furthermore, the effects of various parameters including vehicle wheel position, horizontal load, elasticity, and thickness of the road layers are studied using finite element simulations. Chapter 3 deals with the fracture behavior of HMA (hot mix asphalt) concretes. Different test specimens including the SENB, DC(T), SCB, etc. are employed to measure the fracture resistance of HMA concretes under different modes of loading including pure modes of I, II, and III, and mixed modes of I/II and I/III. The effects of different parameters including aggregate type, aggregate gradation, air void content, binder, temperature, nano-materials, fibers, and additives on the fracture resistance of HMA mixtures are discussed. Chapter 4 discusses the fracture behavior of warm mix asphalt (WMA) concretes. The effects of different parameters including mode of loading, temperature, crumb rubber, fiber, and aggregate type on fracture resistance of WMA mixtures are reviewed. Chapter 5 deals with the application of nonlinear fracture mechanics in asphalt mixtures. Two nonlinear methods including fracture energy and J-integral are usually employed to characterize the fracture resistance of asphalt mixtures. Generally, three different tests including SCB, DC(T), and SENB have been developed to determine the fracture energy. These tests are described in this chapter. Meanwhile, the procedure for determination of critical J-integral J_c, as another nonlinear fracture resistance indicator for evaluating fracture behavior of asphalt mixtures, is discussed in Chap. 5. Finally, Chap. 6 summarized the whole content of the book.

We would like to express sincere thanks to all the researchers, publishers, and journals that authorized the use of some of their copyrighted material herein. We sincerely hope that this book would be useful for pavement, civil, and mechanical engineers and serves as a reference to both graduate and senior undergraduate students.

Finally, both authors wish to thank their families for their patience and support while this book was being prepared.

Sadjad Pirmohammad, Ph.D.
Associate Professor
University of Mohaghegh Ardabili
Ardabil, Iran

Majid Reza Ayatollahi, Ph.D.
Professor and Director
Fatigue and Fracture Research Laboratory
School of Mechanical Engineering
Iran University of Science and Technology
Tehran, Iran

Contents

About the Authors

Dr. Sadjad Pirmohammad received his B.Sc. degree in mechanical engineering from the University of Tehran, and his M.Sc. and Ph.D. degrees from the Iran University of Science and Technology. His Ph.D. thesis was on "study of mixed mode I/II fracture behavior of asphaltic materials at low temperatures." He is currently serving as Associate Professor and Researcher at the University of Mohaghegh Ardabili (Iran) with more than 10 years of experience in research on the fracture behavior of asphalt materials. He has published several books and journal papers on fracture mechanics applications in asphalt concretes.

Dr. Majid Reza Ayatollahi (Ph.D., University of Bristol, UK, 1998) is Distinguished Professor and Director of fatigue and fracture research laboratory in the School of Mechanical Engineering at Iran University of Science and Technology (IUST). His main fields of interest are fracture mechanics, as well as the experimental and computational solid mechanics. He has published more than 250 fully refereed papers in well-known international journals and more than 300 papers in the proceedings of national and international conferences. He is the author of two books both published by Springer International on mechanical behavior of nano-structured materials. He is currently member of the editorial boards in eight international journals and has served as reviewer for more than 50 international journals. He received a prestigious award as "National Distinguished Professor" in the year 2014 from the Iranian ministry of science, research, and technology. He also achieved "National Distinguished Researcher" award from the same ministry in 2017 and several awards from IUST as Eminent Professor/Researcher in different years. He was member of the university research council at IUST for ten years and is currently the Iran's representative in the Asian Society of Experimental Mechanics. He has presented numerous plenary/keynote talks in international conferences and universities.

Nomenclature and Abbreviations

Roman Symbols

a	Crack length (depth)
b	Specimen thickness in Eq. 5.6
A_{lig}	Ligament area
D	Diameter
D, L	Dimensions of surface crack in Fig. 2.22
E	Elasticity (Young's) modulus
E°	Reference elasticity modulus
F_{h}	Horizontal load
F_{v}	Vertical load
G	Energy release rate
G_{f}	Fracture energy
J	J-integral
J_{c}	Critical J-integral
K_{I}	Mode I stress intensity factor
K_{II}	Mode II stress intensity factor
K_{III}	Mode III stress intensity factor
K_{If}	Mode I critical stress intensity factor
K_{IIf}	Mode II critical stress intensity factor
K_{IIIf}	Mode III critical stress intensity factor
K_{eff}	Effective critical stress intensity factor
K^{n}	Normal stiffness in DEM
K^{s}	Shear stiffness in DEM
l	Length
L	Position of crack in SCB specimen
M^{e}	Mixity parameter
n_{Γ}	Unit vector normal to Γ
P	Load
P_{cr}	Fracture load

P_{max}	Maximum value of load
R	Radius
S	Span
S_1, S_2	Positions of the bottom supports in SCB specimen
t	Thickness
t_o	Cohesive strength
T_i	Traction vector
t_n	Traction
T	Temperature
T-stress	Nonsingular component of stress at the crack tip
u	Displacement field
U	Strain energy to failure
w	Width
W_f	Work of fracture
w_s	Strain energy density
Y_I	Mode I geometry factor
Y_{II}	Mode II geometry factor
Y_{III}	Mode III geometry factor

Greek Symbols

α	Crack angle
r, θ, z	Cylindrical coordinates
σ	Stress field
σ_t	Tensile strength
$\sigma_{rr}, \sigma_{r\theta}, \sigma_{\theta\theta}, \sigma_{rz}, \sigma_{\theta z}, \sigma_{zz}$	Stress components in the cylindrical coordinate
ε	Strain field
υ	Poisson's ratio
φ	Location of a point on the crack head shown in Fig. 2.22
μ	Shear modulus
Γ	Arbitrary contour around the crack tip
Γ_o	Cohesive energy
δ_n	Separation in CZM
δ_c	Critical separation in CZM

Abbreviations

2D	Two-dimensional
3D	Three-dimensional
AC	Asphalt concrete
AE	Acoustic emission

AECS	Asymmetric edge cracked semicircular specimen
AECT	Asymmetric edge cracked triangular specimen
CCCD	Center-cracked circular disk
CMA	Cold mix asphalt
CMOD	Crack mouth opening displacement
CNTs	Carbon nanotubes
CR	Crumb rubber
CT	Constant temperature
CZM	Cohesive zone modeling
DBM	Bitumen macadam
DC(T)	Disk-shaped compact tension specimen
DEM	Discrete element method
DIC	Digital image correlation
DTC	Differential thermal contraction
EBA	Ethylene butyl acrylate
EMA	Ethylene methyl acrylate
EPDM	Ethylene propylene diene terpolymer
EVA	Ethylene vinyl acetate
FEM	Finite element method
FPZ	Fracture process zone
HMA	Hot mix asphalt
IDT	Indirect tensile (tension) test
IIR	Isobutene-isoprene copolymer
LEFM	Linear elastic fracture mechanics
LLD	Load line displacement
LPVD	Vertical displacement of the loading point
NMAS	Nominal maximum aggregate size
PE	Polyethylene
PG	Performance grade
PMMA	Polymethylmethacrylate
PPA	Polyphosphoric acid
PS	Polystyrene
PVC	Polyvinyl chloride
RAP	Reclaimed asphalt pavement
RAS	Recycled asphalt shingles
SBR	Styrene-butadiene-rubber
SBS	Styrene-butadiene-styrene
SCB	Semicircular bend specimen
SEBS	Styrene ethylene butylene styrene
SECS	Symmetric edge cracked semicircular specimen
SECT	Symmetric edge cracked triangular specimen
SENB	Single edge notched beam specimen
SGC	Superpave gyratory compactor
SIF	Stress intensity factor
SMA	Stone mastic asphalt

UTM	Universal test machine
VT	Variable temperature
WMA	Warm mix asphalt
XFEM	Extended finite element method

Chapter 1
Cracking in Asphalt Concretes

Abstract Asphalt concretes are made from aggregates, binders, and modifiers. In this chapter, a brief description is given on the ingredients of asphalt concretes. Different modes of deterioration in asphalt concretes including cracking (i.e., fatigue cracking, longitudinal cracking, transverse cracking, block cracking, slippage cracking, reflective cracking, and edge cracking), surface deformation (i.e., rutting, corrugation, shoving, depression, and swell), disintegration (i.e., pothole and patch), and surface defects (i.e., raveling, bleeding, polishing, and delamination) are then introduced.

1.1 Introduction

Flexible pavements are built from several layers of natural granular material covered with an asphalt concrete layer. A flexible pavement bends as it is loaded by a tire. The objective of designing a flexible pavement is to appropriately distribute the applied loads over the road layers for avoiding the excessive bending of any layer.

Pavements have several functions such as [1]

(i) *Provide a smooth riding surface*: for riding convenience, a smooth riding surface (i.e., low roughness) is necessary. Roughness arises from a number of causes such as pavement distress resulting from structural deformation;

(ii) *Provide adequate surface friction* (i.e., *skid resistance*): safety is another requirement for road users. Safety can be related to a loss of surface friction between the road surface and the tire, particularly during wet conditions;

(iii) *Protect the subgrade*: The soil underneath the pavement is called as the subgrade. As it is overstressed by the applied axle loads, it would deform and lose its ability to properly support the axle loads. Hence, the pavement must have suitable strength and thickness to adequately decrease the actual stresses so that they do not exceed the strength of the subgrade;

(iv) *Provide water proofing*: asphalt concrete acts as a waterproofing material to prevent the beneath layers from moisture ingress. Moisture causes the soil to lose its ability for supporting the applied axle loads, which would result in premature failure of the pavement.

© Springer Nature Switzerland AG 2020

S. Pirmohammad and M. R. Ayatollahi, *Fracture Behavior of Asphalt Materials*, Structural Integrity 14, https://doi.org/10.1007/978-3-030-39974-0_1

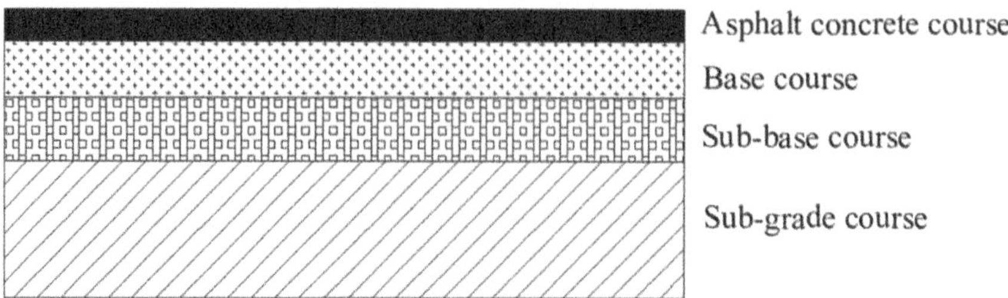

Fig. 1.1 Typical asphalt pavement structure

Asphalt pavements are layered systems with high-quality and expensive materials used in the top layer, where stresses are high, and low-quality and cheap materials used in lower layers. In most asphalt pavements, the stiffness in each layer is greater than that in the layer below and less than that in the layer above [2].

Asphalt pavement structure is typically made up of asphalt concrete course, as top layer, base course, subbase course, and subgrade course, as shown in Fig. 1.1. The purpose of the asphalt concrete course is to provide a smooth and imperfection-free surface [3].

There are various distresses such as rutting, corrugation, shoving, and cracking which influence the performance of asphalt pavements. Among these distresses, cracking is a major threat for the performance of asphalt pavements and imposes significant costs on the road agencies. Good understanding of the cracking mechanisms can be an appropriate way to reduce the costs of maintenance processes. In this chapter, after describing asphalt materials paved on roads, major distresses occurred on the asphalt overlays are briefly explained, and different types of cracking including alligator cracking, longitudinal cracking, transverse cracking, block cracking, slippage cracking, reflective cracking, and edge cracking are introduced.

1.2 Asphalt Concrete

Asphalt concrete is prepared by a combination of two primary ingredients including aggregates and binder (or asphalt cement), and, in some cases, binder modifiers. The aggregates make 90–95% of the total mixture by weight and 75–85% by volume. They are mixed with 5–10% binder to form asphalt concrete. The aggregates and binder are mixed in an efficient manufacturing plant capable of producing specified materials. Asphalt concrete is then transported by truck to the paving site where it is spread to a uniform thickness with a mechanical paving or finishing machine. The asphalt concrete is then compacted to the required degree by heavy and self-propelled rollers to produce a smooth and well-compacted pavement course.

Asphalt concretes are usually classified into three categories: (i) hot mix asphalt (HMA) concrete, (ii) cold mix asphalt (CMA) concrete, and (iii) warm mix asphalt

(WMA) concrete. The HMA concrete properties are much better than the CMA concrete; hence, the HMA concretes are used for higher-volume traffic [4].

In the production of HMA concrete, a blend of aggregates and binders is mixed, transported, placed, and compacted in a heated condition. The binder used in HMA is not liquid at room temperatures; therefore, once HMA concrete cools, it becomes semisolid or solid and strong enough to withstand heavy traffic [4].

On the other hand, CMA concrete is normally handled, placed, and compacted without heating. CMA concrete can be handled cold because it uses liquid binder in the form of an asphalt emulsion. Asphalt emulsion is a blend of asphalt, water, and specific chemical additives. As asphalt emulsion is mixed with aggregate, the asphalt separates from the water (the emulsion breaks) and coats the aggregate. Thus, CMA is a blend of emulsified binder and aggregate, which is produced, placed, and compacted at ambient air temperature. The production of CMA concrete is low cost because it does not need large amounts of energy for heating the asphalt mixture during the production and placement processes. In addition, the cooling of mixture during transportation to the work site is not an issue, as it can be with HMA and WMA concretes. However, in a technical point of view, the CMA is difficult to compact thoroughly, and generally it is not as durable as the temperature-produced mixtures (i.e., HMA and WMA concretes) [4].

In the production of WMA concrete, various methods are employed to significantly decrease mix production temperature by 17 °C to nearly 56 °C. In addition, the rate of cooling of the HMA concrete is higher than that of the WMA concrete, and it is conceivable that both are applied to the pavement construction at nearly the same temperature. The WMA technology has some advantages such as (i) lower cost—significantly less fuel is needed to produce the mixture, (ii) the emissions are lower, hence there is a decreased environmental impact, and (iii) there is the potential for improved performance because of decreased time for hardening (age hardening) [4–6].

The production temperature of WMA concrete is in the range of 100–140 °C, whereas it is in the order of 150–170 °C for HMA concrete. After introducing WMA concrete in 2000, it attracted considerable attention of the highway engineering community because of its benefits over both HMA and CMA concretes. The use of WMA concrete is simple and does not need any major plant modifications to the existing HMA plant system. The manufacturing processes of WMA concrete are different from the HMA concrete [4].

Sasobit [7] is a synthetic wax that is produced in the coal gasification process. It is normally mixed with binders at temperatures above 115 °C. Sasobit has a melting point in the range of 85–115 °C. Furthermore, Sasobit makes a crystalline structure within the binder at ambient temperatures. Therefore, Sasobit decreases the binder viscosity at temperatures above 115 °C and improves the stability at ambient temperatures. The optimum amount of Sasobit used in the WMA concretes is in the order of 3–4% (by weight) of the binder which in turn would lead to decrease the production temperatures by nearly 8–30 °C. It is pointed out that blending the solid Sasobit with the binder during mixing is not recommended, because it will lead to a non-homogeneous distribution of Sasobit within the mixture. Hence, Sasobit is mixed with the hot binder stream to ensure homogeneous distribution. The WMA technology decreases carbon dioxide emissions by 30%, and it consumes 30% less energy and decreases dust emissions by 50–60% compared to conventional HMA production [4].

1.2.1 Aggregates

Aggregates are an important component of the materials used in construction of roads. The physical, mechanical, and chemical properties of aggregates play an important role in the performance of asphalt concretes.

Based on source, the aggregates are classified as natural and artificial. Natural aggregates refer to those derived from the natural rocks; while, artificial aggregates are the by-product of the manufacturing process of other materials. An example of by-product aggregate is slag, which is produced during the metallurgical processing of steel, iron, tin, and copper. The most widely used variety is blast-furnace slag, which is a nonmetallic product that is developed in a molten condition simultaneously with iron in a blast furnace [8].

Aggregates derived from natural rocks can be classified on the basis of size as crushed stone, sand, or gravel. Crushed stone refers to the different rock types and sizes that are produced by blasting and then crushing. Sand and gravel comprise any clean mixture of aggregate sizes found in natural deposits, such as stream channels [8].

Based on the size, the aggregates are classified as fine and coarse. The size that separates fine from coarse aggregates differs, depending on the application and the intended use of the aggregates. For HMAs, the No. 4 (4.75 mm) sieve or the No. 8 (2.36 mm) sieve are typically used to separate the fine aggregate from the coarse aggregate sizes [8].

Aggregate gradation represents the percentage of each of the sizes in a mixture. It is typically expressed as the percentage of the aggregate blend passing sieves with standard openings. The size distribution of aggregate particles directly states the performance of the pavement layers. Generally, aggregate size distributions are classified as gap graded, uniform, well-graded, and open graded. The sieves that are typically used in determining the gradation are 2 in, 1–1/2 in, 1 in, ¾ in, ½ in, 3/8 in, No. 4, No. 8, No. 16, No. 30, No. 50, No. 100, No. 200 (i.e., 50.8, 37.5, 25.4, 19, 12.5, 9.5, 4.75, 2.36, 1.18, 0.6, 0.3, 0.15, and 0.075 mm, respectively). Aggregate gradation is typically shown in a graphical form in which the percent of aggregate passing a sieve size is plotted on the ordinate in an arithmetic scale, and the particle size is plotted on the abscissa in a logarithmic scale [8].

Aggregate blends are designated by their maximum aggregate size or their nominal maximum aggregate size (NMAS). According to ASTM C 125, the maximum size refers to the smallest sieve through which 100% of the aggregate sample particles pass, and the NMAS is defined as the largest sieve that retains some (i.e., less than 10% by weight) of the aggregate particles. The superpave mix design system defines these properties differently. The maximum size is defined as one sieve size larger than the NMAS and the NMAS as one sieve size larger than the first sieve to retain more than 10% by weight [8].

1.2.2 Binder

Asphalt binder is a blend of hydrocarbons of different molecular weights. It is the product of the distillation of crude oil. At the molecular level, binder consists of compounds called hydrocarbons, made of hydrogen and carbon atoms. Binder molecules have (1) an aliphatic structure of straight or branched chains, (2) an unsaturated ring or aromatic structure, or (3) saturated rings of branches, which have the highest hydrogen-to-carbon ratio. Figure 1.2 shows examples of these structures. The atoms within binder molecules are held together by strong covalent bonds [8].

In addition to hydrocarbons, the binder includes heteroatoms such as nitrogen, sulfur, oxygen, and metals. Although these heteroatoms exist in small percentages compared to the hydrocarbons, they influence the interactions among molecules and binder properties. Distribution of metals such as vanadium, nickel, and iron depends on the crude oil source [8].

Different methods have been developed over the years for grading binders. The main objective of these grading systems is to classify binders based on their rheological and mechanical properties, assuming that these properties relate to the field performance. One of the binder grading systems is penetration grading (ASTM D 946) in which binders are graded based on the penetration of a standard needle in binder at 25 °C in units of 0.1 mm. Binders also have to meet other requirements to be graded in this system. There are five grades, i.e., 40/50, 60/70, 85/100, 120/150, and 200/300. The binders are tested in unaged condition. Superpave performance grading is another important grading system. In this system, the tests are conducted

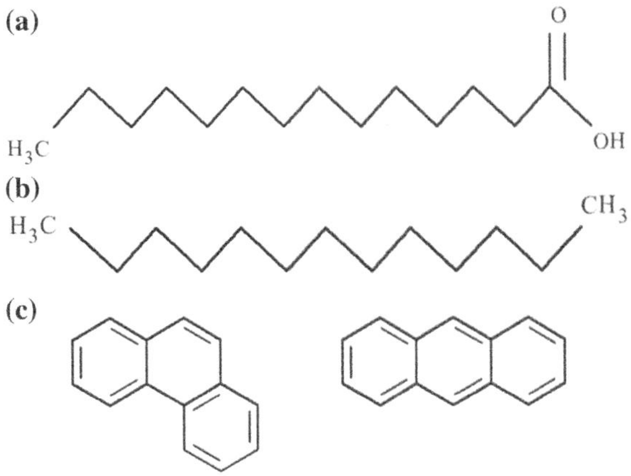

Fig. 1.2 Examples of binder molecules, **a** aliphatic molecule, **b** saturated molecule with branched structure, **c** aromatic molecules [8]

at temperatures that represent the geographic location in which a binder will be used. A number of tests are involved in the superpave system to obtain the binder rheological properties at different temperatures. A binder grade indicates the pavement temperatures at which this binder can be used. For example, PG 64-22 indicates that this binder can be used where the average seven-day maximum pavement temperature is lower than or equal to 64 °C, and the minimum pavement temperature is higher than or equal to −22 °C [8].

It is pointed out that Radovskiy and Teltayev [9] have suggested approximate formulas for prediction of the rheological properties of asphalt binder based on its standard parameters such as penetration and softening point.

1.2.3 Binder Modification

Various materials are employed to improve the binder properties. Binder modification has been driven by the increase in traffic loads, new refining technologies, enhancement in polymer technology, and the increasing need to recycle waste such as rubber. This improvement is realized in enhancing the maximum temperature and/or reducing the minimum temperature at which a binder can be used. According to King et al. [10], the results of modified binder depend on a number of factures such as concentration of the modifiers, molecular weight, chemical composition, particle size, molecular orientation of the additive, crude source, refining process, and grade of the original unmodified binder. Table 1.1 lists some of the binder modifiers used for preparation of asphalt concretes.

1.3   Deteriorations of Asphalt Pavements

Distresses in asphalt pavements are appeared due to the combined effects of traffic loading and environmental conditions. There are four main distresses in asphalt pavement surface, as follows:

(i)   Cracking
(ii)   Surface deformation
(iii)   Disintegration (potholes, etc.)
(iv)   Surface defects (bleeding, etc.).

An overview of these modes of deterioration is presented in Sect. 1.3 mainly based on a very useful article published by Adlinge and Gupta [1].

Table 1.1 Types of binder modifiers [11]

Categories of modifier	Examples
Thermosetting polymers	Epoxy resin
	Polyurethane resin
	Acrylic resin
Elastomeric polymers	Natural rubber
	Volcanized (tire) rubber
	Styrene-butadiene styrene (SBS)
	Styrene-butadiene rubber (SBR)
	Ethylene propylene diene terpolymer (EPDM)
	Isobutene-isoprene copolymer (IIR)
Thermoplastic polymers	Ethylene vinyl acetate (EVA)
	Ethylene methyl acrylate (EMA)
	Ethylene butyl acrylate (EBA)
	Polyethylene (PE)
	Polyvinyl chloride (PVC)
	Polystyrene (PS)
Chemical modifiers	Oregano-manganese/cobalt compound
	Sulfur
	Lignin
Fibers	Cellulose
	Glass fiber
	Asbestos
	Polyester
	Polypropylene
	Kenaf
	Basalt
Antis tripping	Organic
	Amines
	Amides
Natural binders	Trinidad lake asphalt
	Gilsonite
	Rock asphalt
Filler	Carbon blacks
	Fly ash
	Lime
	Hydrated lime

1.3.1 Cracking

The most common types of cracking appeared in the asphalt pavements are as follows:

(i)   Fatigue (or Alligator) cracking
(ii)   Longitudinal cracking
(iii)   Transverse cracking
(iv)   Block cracking
(v)   Slippage cracking
(vi)   Reflective cracking
(vii)   Edge cracking.

1.3.1.1 Fatigue (or Alligator) Cracking

Fatigue cracking, which is usually called as alligator cracking, is a series of interconnected cracks creating small and irregular shaped pieces of pavement. It is generated within the asphalt concrete layer due to repeated traffic loading (i.e., fatigue). This type of cracking finally leads to disintegration of the pavement surface and/or potholes, as illustrated in Fig. 1.3. Furthermore, alligator cracking usually results in drainage problems. Small areas may be repaired by a patch, but larger areas need reclamation or reconstruction. Drainage must be carefully examined in all cases.

1.3.1.2 Longitudinal Cracking

Longitudinal cracks are long cracks that run parallel to the pavement centerline, as shown in Fig. 1.4. Longitudinal cracks are usually caused by frost heaving or joint failures, or they may be load induced. Understanding the cause is essential for choosing an adequate repair. Meanwhile, several parallel cracks may finally form from the initial crack. This phenomenon, known as deterioration, is usually a sign that crack repairs are not the proper solution.

Fig. 1.3 Alligator cracking

Fig. 1.4 Longitudinal cracking

1.3.1.3 Transverse Cracking

Transverse cracks usually occur as asphalt concrete is exposed to cold temperatures before it has completely hardened and while it is still warm. As the temperature reduces, asphalt concrete starts to tighten and then it contracts and shrinks, and this causes it to fracture in a pattern that is perpendicular or transverse to the pavement centerline, as shown in Fig. 1.5. This is why it is also referred to as thermal cracking. Cold temperatures generate the thermal cracking in asphalt concretes, and the moving vehicles worsen it. Meanwhile, transverse cracks will initially be widely spaced (over 20 feet apart). Transverse cracks initiate on the road surface, and then gradually develop deeper and deeper below the road surface if they are not immediately repaired.

Fig. 1.5 Transverse cracking [12]

Fig. 1.6 Block cracking [13]

1.3.1.4   Block Cracking

Block cracking is referred to an interconnected series of cracks that divides the asphalt concrete into irregular pieces, as shown in Fig. 1.6. This may be due to intersecting the transverse and longitudinal cracks. Block cracks can also be due to the lack of compaction during construction. Low severity block cracks can be fixed by a thin wearing course; while, overlays and recycling may be required for the high severity block cracks, and if base problems are found, reclamation or reconstruction may be required.

1.3.1.5   Slippage Cracking

Slippage cracks are half-moon-shaped cracks and are generated in the direction of traffic by the excessive horizontal loads induced from the vehicle braking (see Fig. 1.7). Slippage cracks are usually the result of poor bonding between the asphalt concrete layer and the beneath layer. For example, presence of some materials like

Fig. 1.7 Slippage cracking [13]

Fig. 1.8 Mechanism of reflective cracking

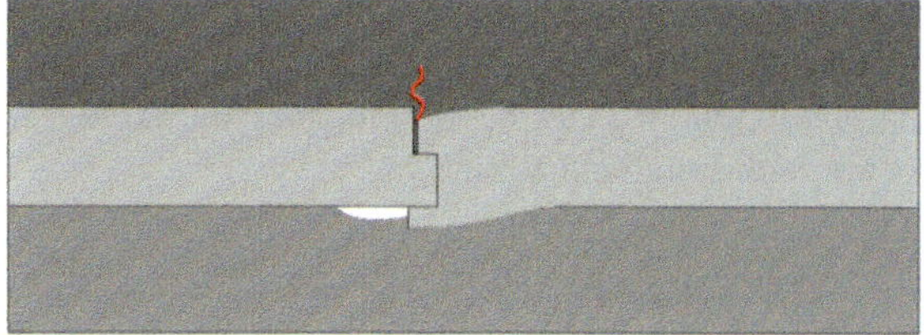

dust, oil, and water between the mentioned layers can result in poor bonding. Repair requires removal of the slipped area and repaving. Meanwhile, a tack coat should be used in the new pavement.

1.3.1.6 Reflective Cracking

When a cracked road is overlaid by a new asphalt concrete layer, cracks under the overlay can cause stress concentration at the bottom of the new asphalt concrete layer, as displayed in Fig. 1.8. Due to the repeated stress concentration, cracks reflect up through the new surface. It is called reflective cracking because it reflects the crack pattern of the pavement structure below. Reflective cracks can also open a way for water to enter the pavement's body and enhance the deterioration rate. Reflective cracks can also occur in asphalt concrete placed on joints or cracks in concrete pavements. It is noticed that before placing any overlays or wearing courses, cracks should be properly fixed.

1.3.1.7 Edge Cracking

Edge cracks form at the edge of the asphalt pavement and in the direction of traffic, as illustrated in Fig. 1.9. These cracks develop from the pavement edge till resembling alligator cracking. Edge cracking results from lack of support of the shoulder because of weak material or excess moisture. They may take place in a curbed part

Fig. 1.9 Edge cracking

of pavement as subsurface water weakens the pavement. At low severity, the cracks may be filled. By enhancing the severity, patches and replacement of distressed areas may be required. In all cases, moisture should be removed, and the shoulders are rebuilt with suitable materials [1].

1.3.2 Surface Deformation

Pavement deformation results from the weakness in road layers due to experiencing movement after construction. The pavement deformation may occur together with cracking. There are various types of surface deformation, as follows:

(i)   Rutting
(ii)   Corrugation
(iii) Shoving
(iv)   Depression
(v)   Swell.

1.3.2.1 Rutting

Rutting is the pavement deformation accumulated in an asphalt pavement surface over time. Rutting is displayed by the wheel path being engraved in the asphalt pavement. Rutting takes place during summer months due to moving the aggregates and binder in asphalt pavement. In very severe rutting, water is held in the rut. Rutting forms in one or more pavement layers, and the width of the rut represents which layer has failed. For example, a very narrow rut states a surface failure, while a wide rut indicates a subgrade failure. Scanty compaction can also result in rutting. Minor surface rutting can be repaired by micropaving or paver-placed surface treatments, and deeper rutting can be smoothed by a truing and leveling course, with an overlay placed over the shim. In addition, reclamation or reconstruction may be required for subgrade failure.

1.3.2.2 Corrugation

Corrugation (or Washboarding) is the formation of periodic and transverse ripples in the pavement surface. Corrugation provides an uncomfortable ride for the vehicle occupants and hazardous driving conditions for vehicles traveling too fast to maintain traction and control. The instability of the asphalt pavement surface maybe because of too much asphalt cement, too much fine aggregate, or rounded or smooth textured coarse aggregate. Minor corrugations can be fixed with an overlay or surface milling. Severe corrugations need a deeper milling before resurfacing.

1.3.2.3 Shoving

Shoving is a plastic movement in the asphalt pavement, generating a localized bulging of the pavement. Locations and causes of shoving are similar to those for corrugations. Minor shoving may be fixed by removing and replacing, and severe shoving can be repaired by milling the surface, followed by an overlay.

1.3.2.4 Depression

Depression is small and localized bowl-shaped area that may have cracking. Water is collected in the depressions. Depressions are usually created by localized consolidation or movement of the supporting layers beneath the asphalt concrete layer due to instability. Depressions can be repaired by excavating and rebuilding. Reconstruction may be needed for severe depressions.

1.3.2.5 Swell

Swell is a localized upward bulge on the pavement surface. Swells form due to an expansion of the beneath layer (i.e., the subgrade). The expansion is usually generated by frost heaving or by moisture. In warmer months, subgrades with highly plastic clays can swell in a manner similar to frost heaves. Swells can be repaired by excavating the inferior subgrade material and rebuilding the removed area. Reconstruction may be needed for severe swelling.

1.3.3 Disintegration

Disintegration is defined as the progressive partitioning of the asphalt concrete into small and loose pieces. Disintegration should be repaired in its early stages; otherwise complete reconstruction of the asphalt concrete may be required. There are two most common types of disintegration, as follows:

(i)   Pothole
(ii) Patch.

1.3.3.1 Pothole

Potholes are a progressive failure and bowl-shaped holes similar to depressions. First, small pieces of the top layer (i.e., asphalt concrete) are dislodged. Over time, the deterioration proceeds into the beneath layers of the pavement. Potholes often form in poorly drained areas and when the pavement disintegrates under traffic loading due to insufficient strength of the pavement layers. Most potholes would not form if the root cause was fixed before the development of the pothole. Potholes can be repaired by excavating and rebuilding. Reconstruction may be needed for extensive potholes.

1.3.3.2   Patch

Patch is a portion of the asphalt concrete that has been removed and replaced. Patches are employed to fix pavement defects or to cover a utility trench. Patch failure can result in failure of the neighboring pavement. A semipermanent patch may be placed for repairing a patch. Area repairs or reclamation may be used for the extensive potholes. Reconstruction is only required for the cases that the base problems are the root source of the potholes.

1.3.4   Surface Defects

Surface defects correspond to problems in the pavement surface layer. The most common types of surface distresses are as follows:

(i)   Raveling
(ii)   Bleeding
(iii)   Polishing
(iv)   Delamination.

1.3.4.1   Raveling

Raveling is the loss of material from the pavement surface. It results from insufficient adhesion between the aggregates and binders in the asphalt concrete. First, fine aggregates break, loose, and leave. By continuing the disintegration, larger aggregates break, loose, and leave, as well. In other words, over a period of time, the thickness of the pavement begins to dwindle away, until the entire pavement is eroded completely. Traffic and freezing weather can accelerate raveling. Raveling can be repaired by a wearing course or an overlay.

1.3.4.2   Bleeding

Bleeding is shiny and black surface film of binder on the pavement surface, which results from the upward movement of binder in the pavement surface. Common causes of bleeding are high binder content in mixture, hot weather, and quality of binder. Excessive binder on the pavement surface reduces the skid resistance of a pavement, and it can become very slippery when wet, creating a safety hazard. Bleeding takes place more often in hot weather conditions when the binder is less viscous, and the traffic forces it toward the pavement surface.

1.3.4.3 Polishing

Polishing is defined as the wearing of aggregates on the pavement surface due to traffic. Polishing can lead to a dangerous low friction surface. A thin wearing course can be used for repairing the pavement surface.

1.3.4.4 Delamination

Delamination is defined as the separation or deboning of pavement layers. Delamination between layers of road is a key pavement failure that can lead to multiple pavement distresses. Asphalt pavement is generally laid in multiple layers of various thicknesses. To act as a single pavement structure, the various layers have to bond together. Failure to bond properly creates problems for the roadway pavement.

1.4 Organization of This Book

This book presents an extensive review of the numerical and experimental studies dealing with the fracture behavior of asphalt concretes. The remaining chapters are organized as follows:

- Chapter 2 deals with the numerical studies related to calculating the crack tip parameters in cracked asphalt overlays and also predicting the fracture behavior of asphalt concretes;
- Chapter 3 discusses on the effect of different material parameters (e.g., binder type, aggregate type, nanomaterials, etc.) on fracture resistance of asphalt concretes;
- Chapter 4 reviews fracture behavior of WMA concretes;
- Chapter 5 describes different test procedures available for characterizing the fracture energy and the critical J-integral of asphalt concretes;
- Chapter 6 summarizes briefly the whole contents of the book.

References

1. Adlinge SS, Gupta A (2013) Pavement deterioration and its causes. Int J Innov Res Dev 2(4):437–450
2. Whiteoak D (1991) The shell bitumen handbook. Surrey, UK, Shell Bitumen

3. Lavin P (2014) Asphalt pavements: a practical guide to design, production and maintenance for engineers and architects. CRC Press, New York
4. Speight JG (2015) Asphalt materials science and technology. Butterworth-Heinemann, Oxford
5. Sargand S, Figueroa JL, Edwards W, Al-Rawashdeh AS (2009) Performance assessment of warm mix asphalt (WMA) pavements. Ohio Research Institute for Transportation and the Environment
6. You Z, Goh S, Dai Q (2011) Laboratory evaluation of warm mix asphalt. Michigan Department of Transportation, Michigan Technological University, Houghton, Final report for project no RC-1556
7. Button JW, Estakhri C, Wimsatt A (2007) A synthesis of warm mix asphalt
8. Papagiannakis AT, Masad EA (2017) Pavement design and materials. Wiley, Hoboken
9. Radovskiy B, Teltayev B (2018) Viscoelastic properties of asphalts based on penetration and softening point. Springer, New York
10. King G, King H, Pavlovich R, Epps AL, Kandhal P (1999) Additives in asphalt. J Assoc Asph Paving Technol 68:32–69
11. Hunter RN (2000) Asphalts in road construction. Thomas Telford, London
12. Pirmohammad S, Zamani H (2018) Technology of asphalt pavement. Moalef Press
13. Ayatollahi MR (2012) Crack propagation in asphalt concretes, Ministry of road and transportation

Chapter 2
Numerical Studies on Asphalt Concretes

Abstract Computational modeling of asphalt mixtures is discussed in this chapter using different methods of DEM (discrete element method), FEM (finite element method) and XFEM (extended finite element method). In the DEM, particles are bonded together at contact points and are separated by external forces. Two homogenous and heterogeneous discrete element models have been generally developed to model crack growth behavior of asphalt concretes. In the homogenous models, behavior of material is considered cohesive at the crack trajectory whilst elastic behavior is assumed at other regions. In the heterogeneous discrete element models, an image processing technique is employed to capture the microstructure of materials. The digital images of specimens having different aggregate sizes are obtained by scanning laboratory asphalt specimens. In the subsequent section, the effects of various parameters including vehicle wheel position, horizontal load, elasticity and thickness of the road layers on the crack tip parameters (i.e. stress intensity factors) are investigated using two-dimensional (2D) and three-dimensional (3D) FEM. For this purpose, a four-layer road structure including asphalt concrete, base, sub-base and sub-grade layers is modeled in ABAQUS. Meanwhile, a top-down crack is regarded within the asphalt concrete layer. Furthermore, extended finite element method (XFEM) has been proved to be a very efficient computational method to characterize the discontinuous mechanical problems such as crack extensions. In the XFEM, the numerical model is divided into two regions. In the first region, the classical finite element meshes are generated for the un-cracked part of geometry; while, in the second region, the meshes defined in the cracked part of geometry are enriched by appropriate functions. Thus, the XFEM incorporates enrichment functions to solve fracture problems. In the last section of this chapter, the simulation of crack propagation using cohesive zone model (CZM) is described. Four material properties including fracture energy, tensile strength (for the cohesive elements), Young's modulus and Poisson's ratio (for the bulk material) together with the load-*CMOD* curve should be experimentally determined.

© Springer Nature Switzerland AG 2020

S. Pirmohammad and M. R. Ayatollahi, *Fracture Behavior of Asphalt Materials*, Structural Integrity 14, https://doi.org/10.1007/978-3-030-39974-0_2

2.1 Introduction

Asphalt concretes are frequently employed as pavement material in road construction due to their outstanding engineering properties such as elasticity, stability, durability, water resistance and noise reduction. In addition to roads, asphalt concretes are sometimes used in parking lots, airports and so on. The life of an asphalt pavement can be extended via good design, construction and maintenance practices.

In general, there are different analytical, numerical and experimental techniques for solving the problems that engineers encounter in real applications [1, 2]. Most of the engineering problems with complex geometry and loading conditions do not have analytical solution. The analytical solution is often limited to ordinary engineering problems. On the other hand, experimental methods are often expensive, time-consuming and laborious. Nowadays, numerical simulation is known as the favorite method to solve engineering problems with complicated geometry, loading and boundary conditions. A numerical simulation implements a mathematical model for an engineering problem. Indeed, numerical simulations investigate the behavior of materials whose mathematical models are too complicated to produce analytical solutions.

Over the past years, different numerical methods including DEM (discrete element method), FEM (finite element method) and XFEM (extended finite element method) have been developed by researchers to simulate fracture behavior of asphalt concretes. In addition, an efficient approach for modeling the initiation and propagation of crack in quasi-brittle materials like asphalt mixtures is cohesive zone modeling (CZM) that takes the evolution of fracture process zone ahead of crack tip into account. This chapter aims at reviewing the applications of these numerical methods on prediction of crack growth behavior of asphalt concretes. Based on these numerical methods, many 2D (two-dimensional) and 3D (three-dimensional) models have been generated to investigate the micromechanical and global fracture behavior of HMA mixtures.

2.2 Fracture Mechanics

Fracture mechanics is a field of mechanical engineering which studies the propagation of cracks in materials. It mainly utilizes both analytical and experimental solid mechanics to determine fracture resistance of materials. Fracture mechanics is an important method for improving crack growth performance of mechanical parts. It employs the stress and strain behavior of materials, particularly the theories of elasticity and plasticity, to the microscopic crystallographic defects found in real materials for predicting the macroscopic mechanical behavior of those parts. The prediction of crack propagation is at the heart of the damage tolerance mechanical design discipline.

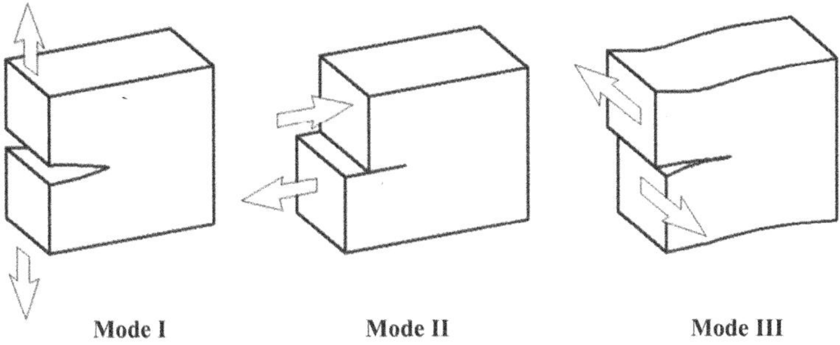

Fig. 2.1 Three major modes of deformation in cracked components

Depending on the relative displacements of crack faces, three major modes of deformation can be defined for a cracked component, as shown in Fig. 2.1 and described below:

- Mode I or opening mode, in which crack faces open without any sliding.
- Mode II or in-plane sliding where crack faces slide relative to each other in the direction of normal to the crack front.
- Mode III or tearing mode, in which crack faces slide relative to each other parallel to the crack front.

There is no crack opening in mode II or mode III. In general, a cracked component subjected to an external load experiences either a pure mode of loading (pure mode I, pure mode II or pure mode III) or a combination of these modes which is called mixed mode loading. When the geometry and loading conditions is complex, cracked structures often undergo mixed mode deformation. This, for example, occurs for top-down or bottom-up cracks which frequently appear in asphalt overlays, when a vehicle wheel passes from its vicinity. More information on the fracture mechanisms of engineering materials can be found in [3, 4].

For a linear elastic material, each stress component around the crack tip can be written as an infinite series expansion in which only the first term is singular. The crack tip stress field under general mixed mode I/II/III conditions can be written as [5]:

$$\sigma_{rr} = \frac{1}{\sqrt{2\pi r}} \cos\frac{\theta}{2} \left[K_{\mathrm{I}}\left(1 + \sin^2\frac{\theta}{2}\right) + \frac{3}{2} K_{\mathrm{II}} \sin\theta - 2K_{\mathrm{II}} \tan\frac{\theta}{2} \right]$$
$$+ T\cos^2\theta + 0\left(r^{1/2}\right) \tag{2.1}$$

$$\sigma_{\theta\theta} = \frac{1}{\sqrt{2\pi r}} \cos\frac{\theta}{2} \left[K_{\mathrm{I}}\cos^2\frac{\theta}{2} - \frac{3}{2} K_{\mathrm{II}} \sin\theta \right] + T\sin^2\theta + 0\left(r^{1/2}\right) \tag{2.2}$$

$$\sigma_{r\theta} = \frac{1}{\sqrt{2\pi r}} \cos\frac{\theta}{2}[K_{\mathrm{I}} \sin\theta + K_{\mathrm{II}}(3\cos\theta - 1)] - T\sin\theta\cos\theta + 0\left(r^{1/2}\right) \tag{2.3}$$

$$\sigma_{\theta z} = \frac{K_{\mathrm{III}}}{\sqrt{2\pi r}} \cos \frac{\theta}{2} + 0(r^{1/2}) \tag{2.4}$$

$$\sigma_{rz} = \frac{K_{\mathrm{III}}}{\sqrt{2\pi r}} \sin \frac{\theta}{2} + 0(r^{1/2}) \tag{2.5}$$

$$\sigma_{zz} = \begin{cases} 0, & \text{Plane stress} \\ \upsilon(\sigma_{\theta\theta} + \sigma_{rr}), & \text{Plane strain} \end{cases} \tag{2.6}$$

where K_{I}, K_{II} and K_{III} are the mode I, mode II and mode III stress intensity factors (SIFs), respectively, v is the Poisson's ratio, and (r, θ, z) are the cylindrical coordinates with their origin at the middle of crack front. The T-term, usually called the T-stress, is a constant and nonsingular term, which is independent of the distance from the crack tip. The higher-order terms of stress field, $0(r^{1/2})$, are often negligible near the crack tip. The crack tip parameters K_{I}, K_{II}, K_{III} and T are functions of the geometry and loading conditions in the cracked specimen. By knowing these parameters, elastic stresses near the crack tip can be calculated from Eqs. (2.1)–(2.6).

2.3 Discrete Element Method

The discrete element method (DEM), also called the distinct element method, is among the numerical techniques used in the past to simulate the mechanical behavior of asphalt layers in the presence or absence of a crack (see for example [6–10]). The method was proposed by Cundall [11] in the context of rock mechanics, and is a powerful numerical tool for simulating the behavior of granular and particulate structures, and also for investigating the micromechanics of materials. This method is utilized in a wide range of disciplines including materials science, process engineering, agricultural engineering, mechanical engineering, etc. The DEM is also used for modeling continuum problems, particularly those that are determined via a transformation from a continuum to discontinuum [12]. According to Cundall and Hart [13], two criteria must be met for a computational model to be regarded as a DEM. They are: (i) finite displacements and rotations including a complete separation are allowed for discrete bodies, and (ii) by proceeding the calculation, the model recognizes new contacts automatically. Several algorithms have been developed for DEM such as the rigid body-spring method (developed by Kawai [14]) and the deformable polygonal discrete element methods (developed by Hocking [15], Bolander and Saito [16], d'Addetta [17]).

In the DEM, particles are bonded together at contact points and are separated by external forces. It models the motion of individual particles as well as the behavior of bulk material [18]. The DEM discretizes a material via rigid elements with simple shape. Each element interacts with neighboring elements based on appropriate interaction laws that are defined at contact points. The analysis includes three computational steps:

- Computation of internal forces, in which the forces at the contact points are calculated.
- Integration of motion equations, in which element displacements are calculated.
- Detection of contact, where new contacts are identified and broken contacts are removed.

In a discrete element analysis, the interaction between the elements behaves as a dynamic process. In the dynamic process, two main laws, namely Newton's second law and force-displacement law, are applied at the contacts. The acceleration of an element which is resulted from the applied forces (including gravitational, external and internal forces) can be obtained by Newton's second law. Velocity and displacement of an element is then calculated by integrating the acceleration. The contact forces are also obtained by the force-displacement law. The equations of motion are integrated in time using the central difference method. More details on DEM can be found in Cundall and Hart [13]. Efficient algorithms, particularly for the internal force evaluations and contact detection, must be employed since the method may be very difficult computationally.

By assuming no damping in the system, the equilibrium equation of DEM can be written as follows:

$$Ma + K\Delta x = \Delta f \tag{2.7}$$

where, M, a and K are the mass matrix, the acceleration vector and the stiffness matrix, respectively. In addition, Δx and Δf are the incremental displacement and force vectors, respectively. The relation between the increments of force and displacement and the increments of moment and rotation are stated by translational and rotational stiffnesses of a particle, as follows:

$$\{\Delta f\} = [K]\{\delta u\} \tag{2.8}$$

For a two-dimensional system, Eq. (2.8) can be written as:

$$\begin{Bmatrix} \Delta F_1 \\ \Delta F_2 \\ \Delta M_3 \end{Bmatrix} = \begin{bmatrix} k_{11} & k_{12} & 0 \\ k_{21} & k_{22} & 0 \\ 0 & 0 & k_{33} \end{bmatrix} \begin{Bmatrix} \Delta U_1 \\ \Delta U_2 \\ \Delta \theta_3 \end{Bmatrix} \tag{2.9}$$

The individual elements of the stiffness matrix can be written in terms of the contact normal vector ni, particle radius R, and the contact stiffnesses (namely K^n and K^s). Hence, Eq. (2.8) can be stated similar to the terms in a standard finite element frame work, as follows:

$$\{\Delta f\} = \left(\int [B]^T[D][B]dV \right)\{\delta u\} \tag{2.10}$$

$$\{\varepsilon\} = [B]\{\delta u\} \tag{2.11}$$

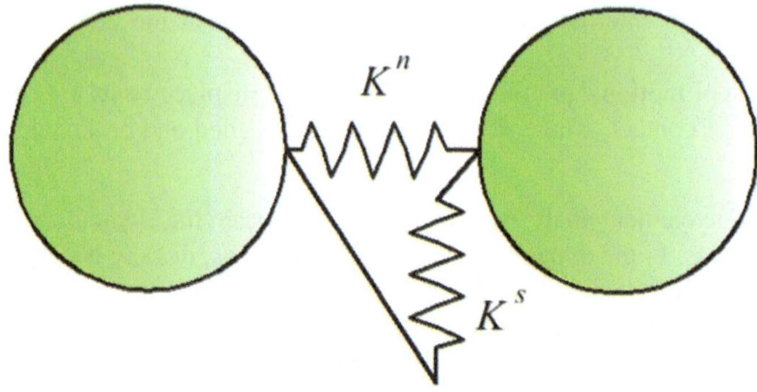

Fig. 2.2 Springs at the contact point [15]

$$\{\sigma\} = [D]\{\varepsilon\} = [D][B]\{\delta u\} \equiv \Delta\sigma_1 \tag{2.12}$$

$$\{\Delta f\} = \int [B]^{\mathrm{T}} \Delta\sigma_1 \mathrm{d}V \tag{2.13}$$

where, $\{\Delta f\}$, $\{[K]\}$, $\{\delta u\}$, $\{\varepsilon\}$, $\{\sigma\}$, $\{[B]\}$ and $\{[D]\}$ express the incremental force vectors, the contact stiffness matrix, the contact displacements, the contact strains, the contact stresses, the strain matrix and the elasticity matrix, respectively.

The components of the stiffness matrix express the normal and shear springs that are available at the contact points (see Fig. 2.2). From the strain energy density, the relationship between the discrete element contact stiffness and elastic properties is defined as [19]:

For plane strain

$$K^{\mathrm{n}} = \frac{E}{\sqrt{3}(1 + \upsilon)(1 - 2\upsilon)} \tag{2.14}$$

$$K^{\mathrm{s}} = \frac{E(1 - 4\upsilon)}{\sqrt{3}(1 + \upsilon)(1 - 2\upsilon)} \tag{2.15}$$

For plane stress

$$K^{\mathrm{n}} = \frac{E}{\sqrt{3}(1 - \upsilon)} \tag{2.16}$$

$$K^{\mathrm{s}} = \frac{E(1 - 3\upsilon)}{\sqrt{3}(1 - \upsilon^2)} \tag{2.17}$$

During the analysis, the stiffness matrix (i.e. $[K]$) changes due to forming and breaking the contacts. Therefore, simulations performed using DEM concept can be known as non-linear dynamic analysis. An analogy may be drawn between the discrete element and finite element frameworks; such that the particles and the contacts

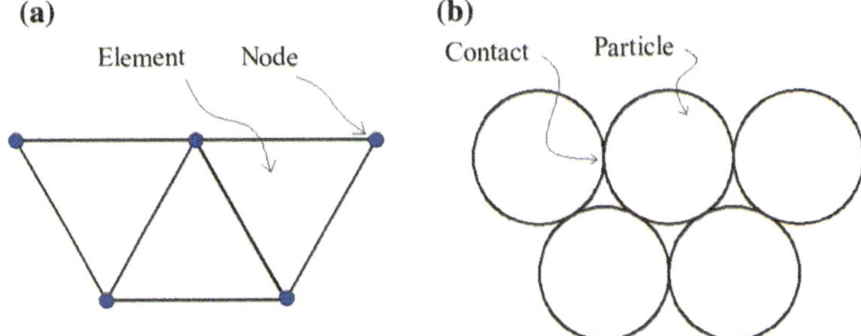

Fig. 2.3 Analogy between the DEM and FEM, **a** elements and nodes (FEM), **b** particles and contacts (DEM) [12]

in DEM correspond to the elements and the nodes in finite element method (FEM), as displayed in Fig. 2.3 [12].

In the next section, some practical examples of DEM applications in simulating the asphalt concrete behavior will be discussed.

2.4 Micromechanical Fracture Models

The progress of micromechanical models backs to over hundred years ago. Micromechanical models are able to predict global material behavior based upon the properties of the individual constituents which have been developed from 1880s. Generally, micromechanical models can be classified into two categories: non-interacting particles and interacting particles. As a further sub-classification, non-interacting particle models can have specified geometries or non-specified geometries [12].

In asphalt concretes, there are several parameters required to be regarded for describing the system which includes one or more disperse phases embedded in a continuous matrix. These parameters are: (i) particle shape, (ii) size and size distribution of particles, (iii) concentration and concentration distribution of particles, (iv) orientation of particles, (v) spatial distribution of particles, (vi) composition of disperse phase, (vii) composition of continuous phase, and (viii) bond between the continuous and disperse phases. These parameters ought to be initially incorporated into the analyses. After reaching to a satisfied grasp, homogenization techniques at considered length scales can be utilized to simplify the final model. The need to consider interacting particles for the precise prediction of asphalt concrete stiffness properties was demonstrated from 1990s. The complex morphological features at the meso-scale, namely the contact of irregularly shaped aggregates require numerical methods to account for the particle structure. This has been addressed using various numerical methods such as finite element method [20] and discrete element method [17].

Asphalt concrete is considered as a quasi-brittle material composed of aggregates and binder. The fracture of heterogeneous materials is a complex problem because of the creation and movement of new cracked surfaces. The application of nonlinear fracture mechanics to describe fracture mechanisms is an extreme challenge, since the fracture patterns comprise a main crack, crack branches, secondary cracks and micro-cracks. More realistic fracture models with the heterogeneous microstructure have been recently considered using the discrete element method. For example, Kim and Buttlar [21] studied fracture mechanisms of asphalt concretes by developing a heterogeneous fracture model based on the DEM. Two asphalt concrete mixtures with NMASs (nominal maximum aggregate size) of 19 and 9.5 mm were used in this investigation. Both of which used the same aggregate type (i.e. dolomite limestone) and binder (i.e. 5.25% PG64-22). Different tests (i.e. uniaxial complex modulus tests and indirect tensile (IDT) tests) were carried out to find the bulk material properties of aggregates, mastics, and asphalt concretes. Elastic modulus of asphalt concrete mixtures was determined at a temperature condition of $-10\,°C$ with 10 Hz loading frequency, using a compressive uniaxial dynamic modulus test (ASTM D3497-79, 2003). The tensile strength was determined by following the AASHTO T322 (2004) test procedure, with average value of 3.58 (for the mixture of NMAS = 9.5 mm) and 3.20 MPa (for the mixture of NMAS = 19 mm).

Since there is no approved standard test procedure for measuring fracture energy of asphalt concretes, a disk-shaped compact tension (DC(T)) specimen was employed for this purpose. Geometry and dimensions of the DC(T) specimen are shown in Fig. 2.4. The test was performed with a constant *CMOD* (which is monitored by a clip gage) rate of 1 mm/min.

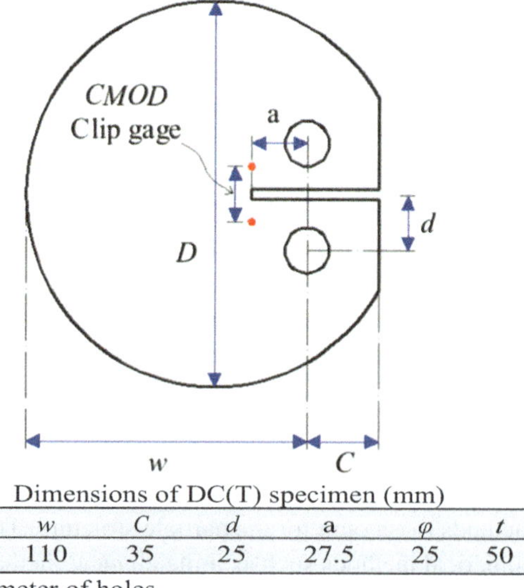

Fig. 2.4 Geometry and dimensions of DC(T) specimen used in [21] for determination of fracture energy

Dimensions of DC(T) specimen (mm)

D	w	C	d	a	φ	t
150	110	35	25	27.5	25	50

φ: Diameter of holes
t: Thickness of the DC(t)

An image processing technique was used for capturing the microstructure of asphalt mixtures. Digitalized specimen image with different aggregate size can be obtained by scanning laboratory asphalt specimens. For instance, images of the specimen with aggregate size ranging from 0.6 to 19, 1.18 to 19, and 2.36 to 19 mm are shown in Fig. 2.5. The digitized images were then projected onto a discrete element mesh, as illustrated in Fig. 2.6. It is reminded that the image with aggregate size ranging from 1.18 to 19 mm was used for the numerical simulations. Different particle arrangements were used for meshing in the DEM such as those illustrated in Fig. 2.7. As shown in Fig. 2.6, a horizontal hexagonal particle arrangement was used in the investigation made by Kim and Buttlar [21]. The hexagonal arrangement is the densest packing method, and thus it was preferred to other arrangements. Structure of

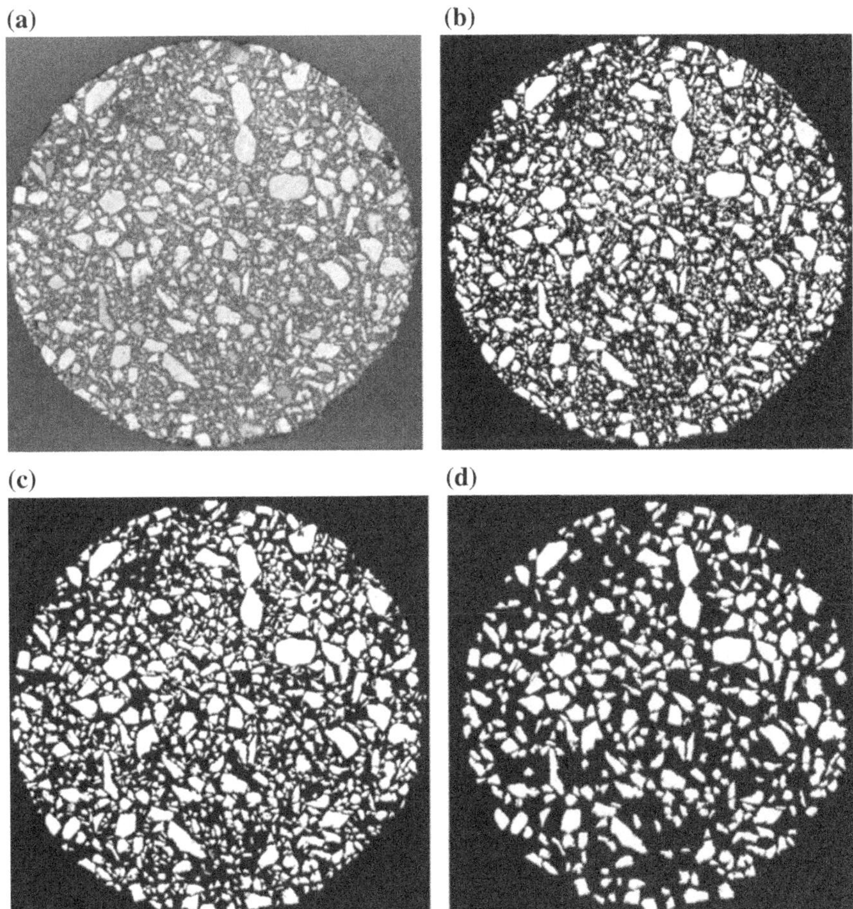

Fig. 2.5 a Laboratory specimen, and digitized images with aggregate size ranging from, **b** 0.6–19 mm, **c** 1.18–19 mm, **d** 2.36–19 mm [21]

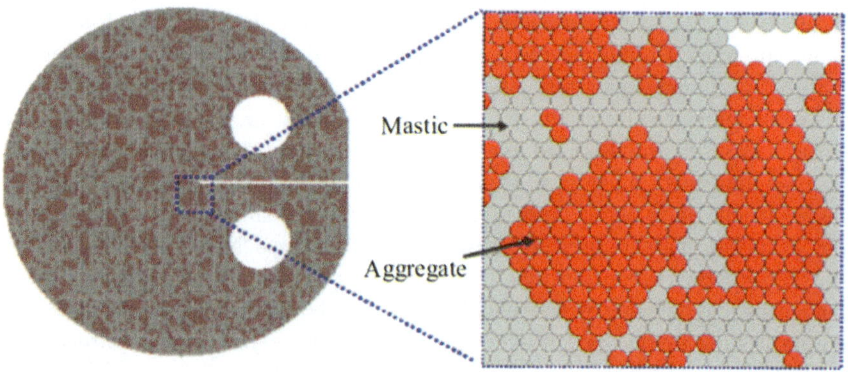

Fig. 2.6 Specimen mesh with hexagonal particle arrangement for the mixture with NMASs of 19 mm [21]

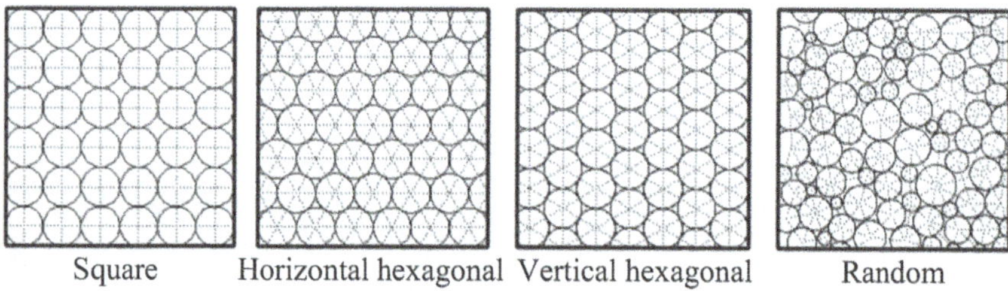

Fig. 2.7 Particle arrangements for meshing in the DEM [19]

asphalt concrete mixtures is made up of three different phases including aggregates, mastic (binder and fine sands), and the interface between aggregates and mastic. According to Fig. 2.8, material properties (i.e. elastic modulus, tensile strength and fracture energy) of these phases were required for numerical simulations, and must be determined separately. Aggregates were assumed to behave in linear elastic without softening. Mastic and interface were assumed to behave in nonlinear elastic with cohesive softening and adhesive softening, respectively. Some of the material properties are difficult to be determined due to experimental restrictions like the interface parameters. The material properties used by Kim and Buttlar [21] in the DEM models are presented in Table 2.1.

A total of 38,721 particles with radii of 0.35 mm and 114,994 contacts were defined for the DEM fracture model displayed in Fig. 2.6. From these numbers, 15,714 particles and 32,738 contacts were considered for the aggregates, 23,007 particles and 54,008 contacts for the mastic, and 28,248 contacts for the interfaces. For the DEM specimen of 9.5 mm NMAS, a total of 38,684 particles with radii of 0.35 mm and 114,857 contacts were used. For this DEM model, 13,126 particles and 25,560 contacts were defined for the aggregates, 25,558 particles and 62,116 contacts for the mastic, and 27,181 contacts for the interfaces. According to the results

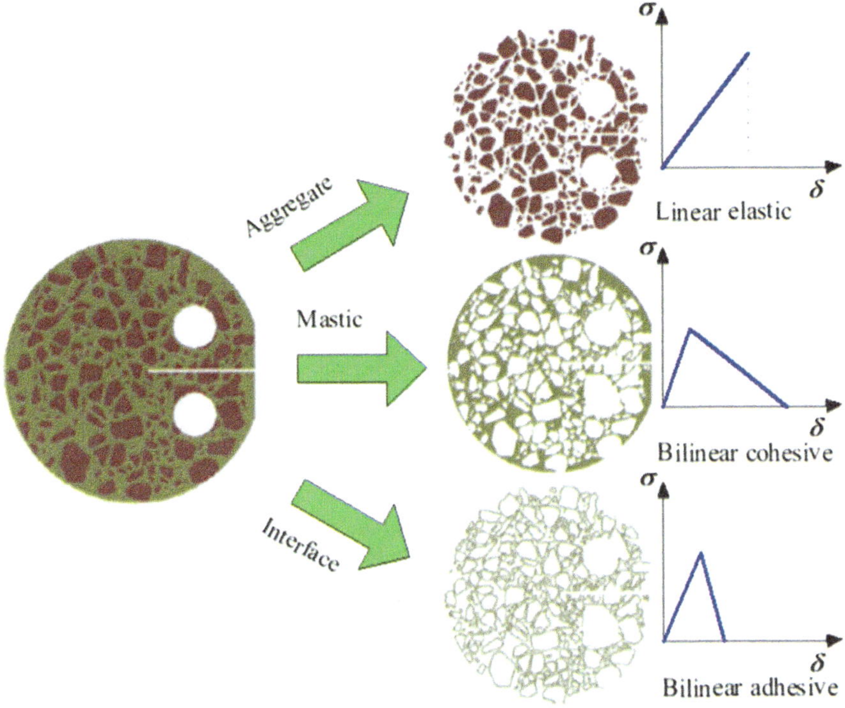

Fig. 2.8 Material models used in the DEM simulation [21]

Table 2.1 Material properties for the mixtures with NMASs of 9.5 and 19 mm conducted at $-10\,^{\circ}\mathrm{C}$ [21]

Material properties					Properties used in DEM		
NMAS (mm)	Poisson's ratio	Phase	Elastic modulus (GPa)	Tensile strength (MPa)	K^{n} (GN/m)	K^{s} (GN/m)	Bond strength (N)
19	Aggregate	0.15	56.8	6.59	2.04	0.81	133.24
	Mastic	0.25	11.4	2.87	0.52	0.0021	58.05
	Interface	0.25	11.4	2.61	0.52	0.0021	52.87
9.5	Aggregate	0.15	56.8	6.59	2.04	0.81	138.57
	Mastic	0.25	18.2	3.78	0.84	0.0034	79.53
	Interface	0.25	18.2	3.44	0.84	0.0034	72.44

presented in [21], generally, there was a good consistency between the experimental and numerical results. However, the numerical results were slightly different from the experimental ones, although the DEM models had the same microstructures as laboratory specimens. Some reasons have been suggested in [21] for this difference such as: (i) the real crack path was not 2D, (ii) aggregates smaller than 1.18 mm and air voids were not taken into account in the simulations, (iii) exact symmetric loading conditions could not be produced during laboratory tests, while it was possible in the simulations.

In another study, Kim et al. [12] studied fracture behavior of asphalt concrete by directly accounting for the contribution of the material's heterogeneity using the DEM. They implemented a clustered distinct element modeling approach in the two dimensional particle flow software package (PFC-2D) for investigating the complicated fracture behavior of asphalt concrete. Furthermore, they introduced a powerful integration of experimental test and numerical scheme, by using the cohesive softening model and image analysis, to simulate crack nucleation, initiation and propagation under mode I and mixed-mode I/II loading.

Two different types of experiments must be carried out to determine fracture energy and tensile strength of asphalt concrete for numerical simulations by the DEM. In order to develop the fracture tests and therefore to find the fracture energy, a single edge notched beam (SENB) was selected as the test geometry. The SENB specimens were fabricated from laboratory compacted beams or slabs. It is also noticed that a typical asphalt concrete paved in central Illinois was used to produce the SENB samples. This asphalt concrete consisted of a 9.5 mm NMAS and a binder with performance grade of 64-22. Figure 2.9 shows dimensions of the produced SENB specimen. Closed-loop servo-hydraulic equipment was used to load the SENB specimens using three-point bend fixture. In this experiment, the crack mouth opening displacement (*CMOD*) was increased at a linear rate. The fracture energy of 344 J/m^2 was then determined by calculating the area under the load-displacement curve obtained from the SENB tests and normalizing by the cross-sectional area of the SENB specimen. Since the fracture energy calculated from the SENB tests overestimates the energy related to the material separation; therefore, this parameter must be adjusted for model calibration.

As mentioned above, another important parameter required for numerical analysis is the tensile strength which was determined using the indirect tension test (IDT) conducted at -10 °C and 1 Hz. The procedure for characterization of the tensile

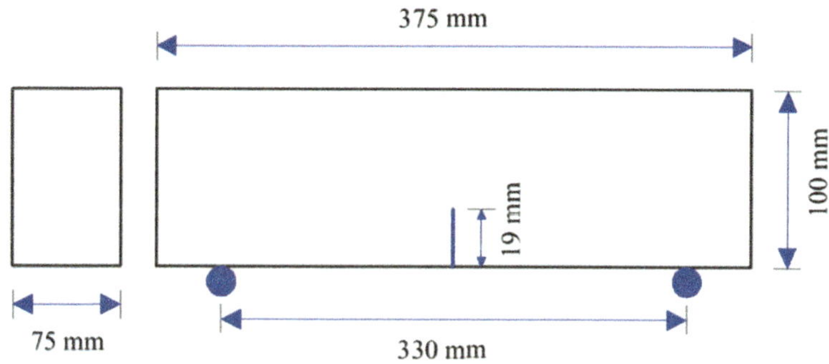

Fig. 2.9 Dimensions of the produced SENB specimen [12]

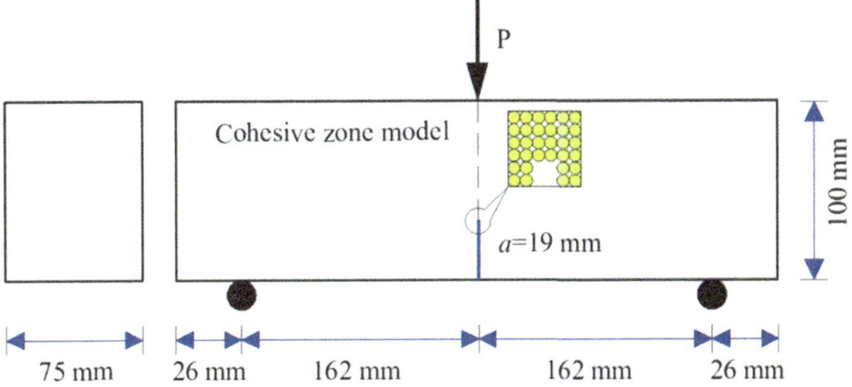

Fig. 2.10 Fracture model geometry and mesh using DEM under mode I [12]

strength of asphalt concretes has been described in the AASHTO T322-03 specification. According to [12], the tensile strength was measured 3.56 MPa using the IDT test.

The SENB geometry with the aforementioned dimensions was modeled numerically by using the DEM. Figure 2.10 illustrates a discrete element configuration around crack tip. Meanwhile, cohesive contact models were considered only along the crack line. The bulk material was regarded as elastic, homogeneous, and isotropic. The experimentally obtained material properties (i.e. strength and fracture energy) were used as the material inputs into the cohesive fracture model. Both the homogenous and heterogeneous models were employed in [12] to investigate fracture mechanisms of asphalt concrete.

In the homogenous model, they used 37,981 particles with 0.5 mm radius and 75,442 contacts across the entire specimen, while 81 cohesive contacts were considered along the crack line. Elastic modulus was taken 14.2 GPa for the bulk material. It is worth noting that the calibrated tensile strength of 3.21 MPa and the calibrated fracture energy of 241 J/m^2 were assigned for cohesive fracture model. They conducted a parametric study on the effect of the particle size on the numerical results in the homogenous model. Three different particle radii (0.25, 0.5 and 1.0 mm) were regarded in this study, and the results demonstrated that particle size did not affect the global fracture response [19]. However, particle size was an important aspect in the heterogeneous fracture model; because, it affected distribution of micro-cracks [22]. Based on the experimental and numerical results, the force-*CMOD* curve obtained from the calibrated model was quite reasonable in comparison to the experimental results.

Kim et al. [12] also constructed a heterogeneous discrete element fracture model, on the SENB geometry shown in Fig. 2.10 under mode I loading, by using high resolution optical image equipment and powerful image processing techniques. The realistic microstructure was obtained and projected onto the discrete element mesh. Geometry of the heterogeneous model and crack growth results are plotted in

Fig. 2.11. In this analysis, 149,922 particles with 0.25 mm radius and 298,855 contacts were utilized for the entire SENB specimen. From these numbers of particles and contacts, 61,717 particles and 105,236 contacts were used to model aggregates; 88,205 particles and 158,042 contacts for mastic, and 35,577 contacts for interfaces between aggregate and mastic. Values of the parameters used in the numerical model have been listed in Table 2.2. According to Fig. 2.11a, many micro-cracks were predicted in the process zone before crack extension. Because of the non-homogeneity aspects, asphalt materials have many different fracture mechanisms like micro-cracking, crack branching and deflection, crack face sliding, crack bridging, and crack tip blunting. A part of the external energy caused by the applied load is consumed by the micro-cracks formation. Crack deflection takes place when the path of least resistance is around a relatively strong particle or along a weak interface. In addition, frictional sliding between the cracked faces during the opening of

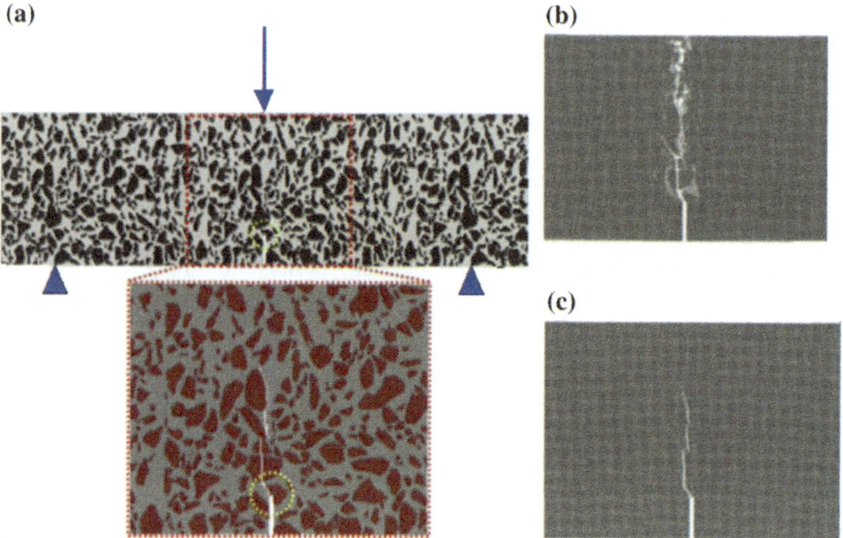

Fig. 2.11 Heterogeneous DEM fracture model, **a** microstructure of SENB, **b** micro-cracks, **c** macro-cracks [12]

Table 2.2 Parameters used in heterogeneous DEM fracture model [12]

Phase	Material properties		DEM contact properties		
	Elastic modulus (GPa)	Strength (MPa)	Stiffness (GPa)	Bond force (N)	Separation displacement (m)
Aggregate	56.8	6.59	4.26	247.2	7.6×10^{-8}
Mastic	18.2	3.78	1.36	141.75	1.4×10^{-4}
Interface	18.2	3.44	1.36	136.18	4.5×10^{-5}

a tortuous crack causes energy dissipation. The force-*CMOD* curve obtained from the calibrated heterogeneous fracture simulation was similar to that obtained from the experiment. Because of the presence of a large aggregate ahead of the notch, the force experienced a sudden reduction before the peak load but then was recovered.

In addition to the heterogeneous mode I fracture model, Kim et al. [12] constructed the heterogeneous mixed mode I/II fracture model using the mentioned SENB geometry and the discrete element concept. Mixed mode I/II can be achieved by moving the initial crack away from the middle of the specimen. The mixed-mode fracture in a road is popular and important. They studied this issue by using the calibrated cohesive parameters and particle size 0.25 mm. In this investigation, the offset parameter, γ, was taken 0.4, 0.5, and 0.55 resulting in offset lengths of 65, 81, and 89 mm from the middle of specimen (see Fig. 2.12). For example, Fig. 2.13 illustrates the progressive crack initiation and growth that occurred in the numerical simulations of SENB specimen with $\gamma = 0.4$. In the SENB specimen, the mode II proportion at the crack tip increased while the mode I proportion decreased as the crack moved away from the centerline of SENB.

Gao et al. [23] carried out a set of experiments on the SENB specimens made of asphalt mixture with different aggregate distributions and different notch locations.

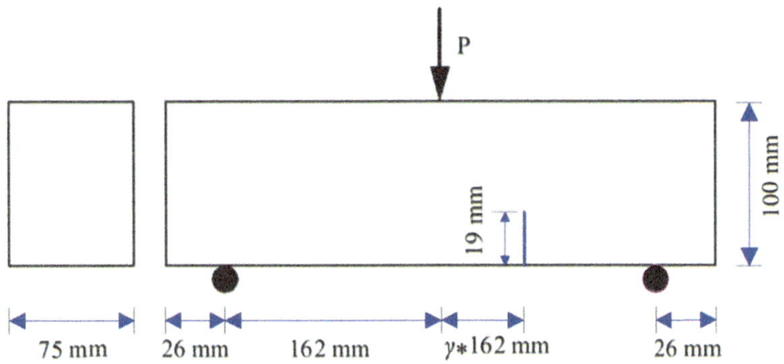

Fig. 2.12 Geometry of SENB loaded under mixed mode I/II [12]

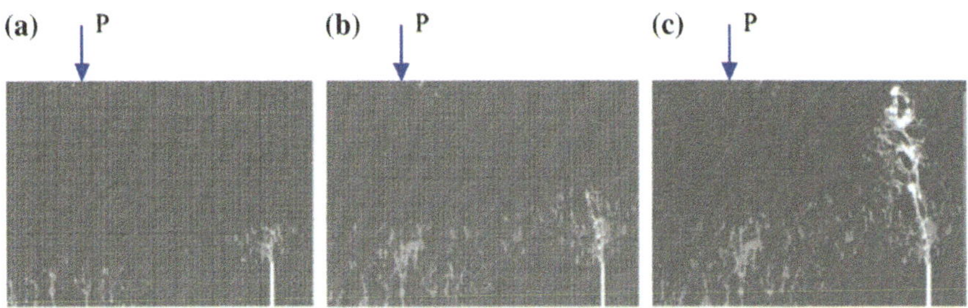

Fig. 2.13 Progressive crack initiation and growth occurred in the numerical simulations of SENB with $\gamma = 0.4$, **a** micro-cracks initiation, **b** macro-crack initiation, **c** crack growth [12]

The numerical image processing technique was then utilized to capture the coarse aggregate information (such as aggregate shape, gradation and distribution) in the SENB specimens, and was combined with the DEM to simulate the fracture behavior. In this study, fine aggregates smaller than 2.36 mm and binder were assumed as the mastic; hence, asphalt concrete was considered as a two-phase composite made up of coarse aggregates with sizes larger than 2.36 mm and homogenized mastic.

The SENB specimens had length of 200 mm, height of 50 mm and thickness of 40 mm. The aggregates size ranging from 0.075 to 16 mm was used for the specimen preparation. A 10 mm notch was respectively created in the middle (i.e. site 1) or 20 and 40 mm off-center locations (i.e. sites 2 and 3) of each specimen. The experiments were carried out by a three-point bend fixture at 5 °C. The two bottom supports were 180 mm away from each other, and the load was applied by a fixture located at the center of SENB with displacement rate of 5 mm/min. Aggregates were considered as a linear elastic material with the elastic modulus of 45 GPa, Poisson's ratio of 0.15 and density of 2500 kg/m^3. The mastic nearly behaves in a linear elastic material with elastic modulus of 1.38 GPa, Poisson's ratio of 0.25 and density of 2200 kg/m^3. Normal stiffness, shear stiffness and bond strength used in [23] were 1300 MN/m, 510 MN/m and 692.50 N (for aggregate), 51 MN/m, 0.21 MN/m and 397.65 N (for mastic), 51 MN/m, 0.21 MN/m and 362.20 N (for interface).

Fracture tests on the SENB specimens containing a notch located at the site 1 (pure mode I loading) with different aggregate distributions shown in Fig. 2.14 were performed to assess the effect of aggregate distributions including distributions 1, 2 and 3. Two characteristics are observed from Fig. 2.14: (1) all the cracks grow only inside the mastic or through the interfaces between the aggregates and mastic due to having much lower fracture resistance than the aggregates; (2) the cracks grow along the crack line although their paths are rough and tortuous due to the obstructions from coarse aggregates. When a crack encounters a coarse aggregate, it bypasses this coarse aggregate; because, aggregates are usually stronger than both the mastic and interfaces. Nevertheless, fracture of the largest aggregate shown in Fig. 2.14b is an exception. It may be because this aggregate has too large size to be bypassed by the crack. Accordingly, aggregate distribution can affect the fracture behavior of asphalt concretes significantly.

The SENB specimen described above was then analyzed via the DEM. A total of 11,600 particles with horizontal hexagonal packing arrangement were used to simulate the fracture tests. These particles had a constant diameter of 1 mm. Figure 2.15 exhibits the crack paths for the SENB specimens with different aggregate distributions 1, 2, and 3. These paths were nearly similar to those displayed in Fig. 2.14 except that the crack did not break the largest aggregate shown in Fig. 2.14b.

In the next step, the effect of pre-notch location (or mixed mode I/II loading) on the fracture behavior was investigated. Figure 2.16 shows the crack trajectories obtained from the numerical simulations. As seen in Fig. 2.16, the cracks propagated inside the mastic or along the interfaces, and also the crack trajectories had the trends pointing to the upper loading point. There was obviously good agreement between the experimental and numerical results for the notch located at the site 2 (shown in Fig. 2.16a), but large difference between those for the notch located at the site 3

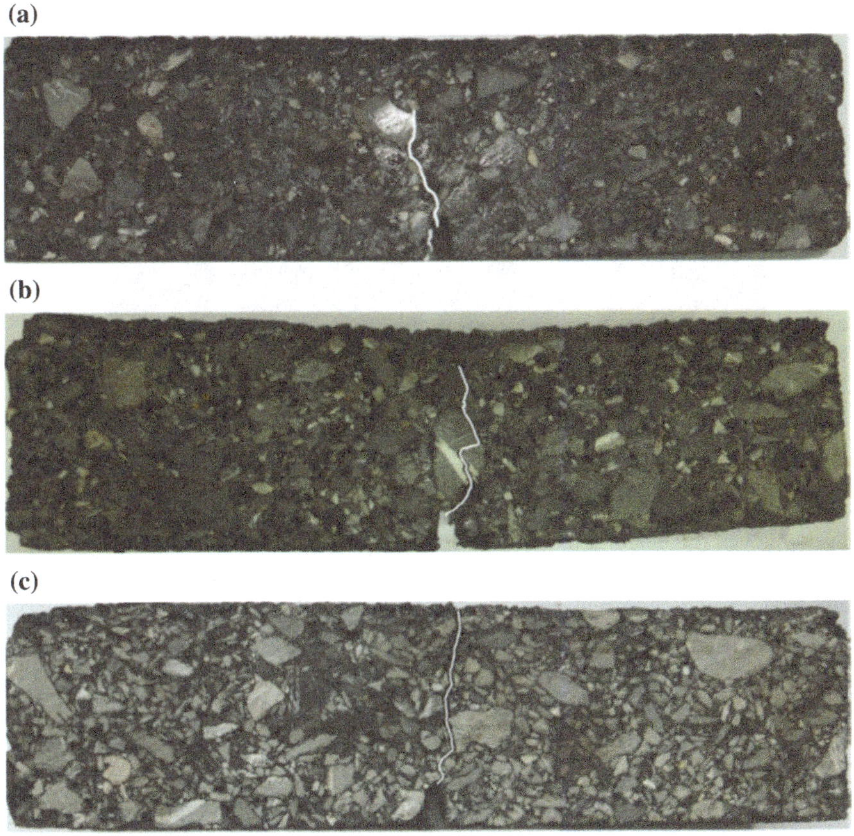

(a)

(b)

(c)

Fig. 2.14 Crack paths in the SENB specimens under pure mode I with different aggregate distributions, **a** distribution 1, **b** distribution 2, **c** distribution 3 [23]

(a)   **(b)**   **(c)**

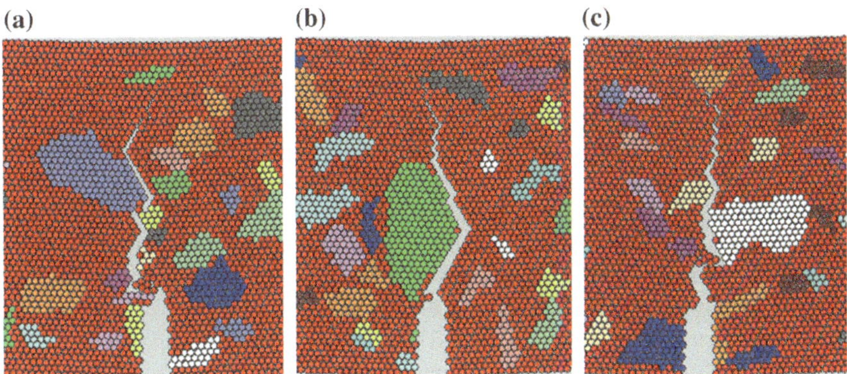

Fig. 2.15 Crack paths in the SENB geometries containing a notch located at the site 1 (pure mode I loading) with different aggregate distributions i.e. **a** distribution 1, **b** distribution 2, **c** distribution 3 [23]

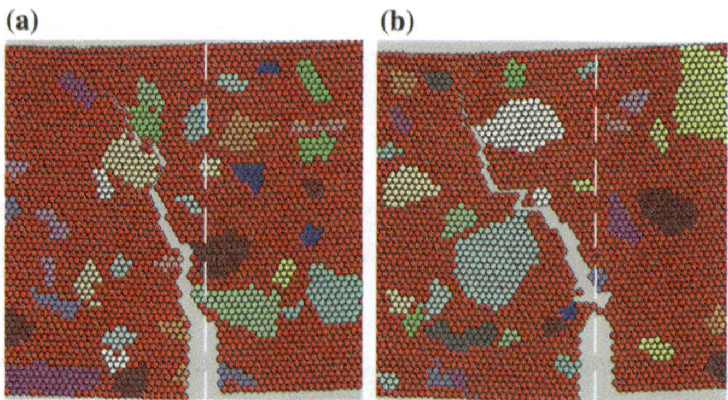

Fig. 2.16 Crack trajectories in the SENB geometries with a notch located at the, **a** site 2, **b** site 3 (mixed mode I/II loading) [23]

(shown in Fig. 2.16b). For the SENB geometry shown in Fig. 2.16b, when the crack traveled 2/3 of the beam height, it ran into a large aggregate. The crack propagated towards the right side and left site of this aggregate respectively in the experiment and numerical simulation. A real crack trajectory was usually not only affected by coarse aggregates distributed in the SENB surface but also by those in the inner. Hence, the abovementioned difference between the experiment and numerical simulation was related to this fact that the effect of the aggregates available inside the SENB specimen could not be regarded in a two-dimensional simulation.

2.5 Finite Element Modeling of Road Structures

Many efforts have been made to minimize the cracking deteriorations in asphalt overlays. As a crack takes place within the asphalt concrete layer of a road, the infiltration of water can lead to rapid degeneration of the road foundation. Therefore, there is a great need to understand the cracking mechanisms for diminishing road degenerations.

In terms of the propagation mechanism, cracks can be classified into fatigue cracks, reflective cracks (bottom-up) and top-down cracks. Many investigations have been performed on the fatigue cracking (see e.g. [24, 25]) and the reflective cracking (see e.g. [26–28]). These cracks are considered as bottom-up cracks signifying that the crack initiates from the bottom of asphalt concrete layer and then propagates upward to the road surface. Conversely, the top-down crack initiates from the road surface and then propagates downward. There are many causes responsible for the top-down cracking such as traffic loads, environmental factor (for example pavement aging), thermal stresses due to extreme cooling rates [29, 30], etc. In particular, when a crack exists on the pavement surface, vehicle traffic can contribute in the crack propagation through the asphalt concrete thickness.

Many numerical researches, including two-dimensional (2D) and three-dimensional (3D), have been performed in the past decades on the road structures; for example, Scarpas et al. [31] used two-dimensional FEM to obtain fatigue life of an asphalt overlay containing bottom-up reflective crack and deboning crack (available in the interface of new and old asphalt concrete overlays). They estimated the number of fatigue cycles using the Paris law. Zhou et al. [32] performed 2D finite element analyses to compute crack parameters (stress intensity factors (SIFs)) at the tip of reflective crack within the road structure. Loizos et al. [33] conducted 2D finite element analyses on a road structure containing a top-down crack. They calculated the SIFs at the crack tip by regarding a moving force. Luo et al. [34] simulated a cracked road structure, and concluded that the crack growth rate increased as the overlay stiffness enhanced. In another study, Hyunwook et al. [35] investigated reflective cracking in the airfield pavement using two-dimensional FE analyses, and obtained SIFs to explore if crack growth occurs by heavy airplane passages. Yang et al. [36] assumed a reflective crack within asphalt concrete to obtain the dynamic SIFs. Their results demonstrated that the longer the initial crack, the less the times of sustaining load of the same extending length. The increase of asphalt overlay thickness can reduce the value of dynamic stress intensity factor to a certain extent. With the augmentation of asphalt overlay modulus, the effect of dynamic load on reflective crack can be reduced in some degree. Dynamic stress intensity factor increased with the increase of joint width. Ceylan et al. [37] conducted many finite element analyses to calculate the SIFs at the tip of the reflection crack for a wide variety of crack lengths and pavement structures. In their study, the neural networks methodology was successfully used to model the SIF as cracks grew upward through a HMA overlay as a result of both load and thermal effects with and without reinforcing interlayer. The effect of thermal stresses on mode I fracture behavior of asphalt concrete has been investigated by Shalaby et al. [38] using 2D and 3D finite element analyses. Suo [39] employed the fatigue destructive mechanics (fracture mechanics and damage mechanics) to develop a 3D finite element model for predicting long-term performance of asphalt concrete pavements. Garzon et al. [26] simulated an airport road structure containing a reflective crack to study the effect of wheel load on the initiation and propagation of crack. Su et al. [40] investigated the effect of tire pressure on the shear stresses. In another study, Ameri et al. [41] investigated the effect of traffic load on the SIFs of a top-down crack located in the middle of the asphalt concrete overlay using 3D finite element simulations.

A brief review concerning the numerical simulations conducted on the road structure was given in the above paragraphs. Details of a number of researches performed in recent years on the cracked road structure are explained in the subsequent sections.

2.5.1 Parametric Study on Road Structure: 2D Investigations

Pirmohammad and Majd-Shokorlou [42] carried out a parametric study on a four layered road structure containing a top-down crack using 2D finite element analysis. In this investigation, the effect of different parameters including vehicle wheel position, horizontal load, elasticity modulus and thickness of the road layers was studied. The road structure involved four layers of hot mix asphalt (HMA) as a top layer, base, sub-base, and sub-grade. As seen in Fig. 2.17, a top-down crack of length $a = 7$ cm was assumed within the HMA layer. The road length and vertical load inducing from a vehicle wheel were assumed as 1200 cm and 40 kN, respectively. In addition to the vertical load (i.e. F_v), the road was loaded horizontally (i.e. F_h) as well. Imprint of a wheel on the road was assumed to be 26 cm, and therefore the vertical and horizontal loads were applied on this segment of road, as shown in Fig. 2.17. It is worth noting that the horizontal loads result from the vehicle acceleration or deceleration (i.e. braking). The position of the load (or vehicle wheel) was described by the parameter L. As a vehicle wheel located in the left side of the crack, the sign of L was considered as negative; while, its sign was positive as the vehicle wheel located in the right side of the crack. It is also evident that $L = 0$ refers to the case that the vehicle wheel was exactly located upon the crack.

Road layers (i.e. asphalt concrete, base, sub-base and sub-grade) are often considered as homogeneous and isotropic, and their behavior is also assumed as elastic [43]. These assumptions work very well in the numerical analyses (see e.g. [44, 45]).

Mechanical properties and thicknesses of the road layers investigated in [42] are presented in Table 2.3. In this research study, the cracked road was modeled in ABAQUS using 8-node elements, and singular elements with nodes at quarter-point positions were used for the first ring of the elements around the crack tip.

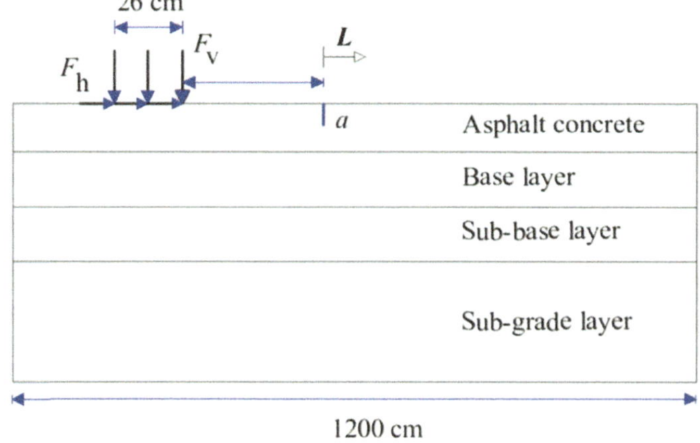

Fig. 2.17 Cracked road structure subjected to vertical and horizontal loads [42]

Table 2.3 Mechanical properties and thickness of the road layers [46]

Layer type	Young's modulus, $E°$ (MPa)	Poisson's ratio, υ	Thickness, t (cm)
HMA	2760	0.35	14
Base	276	0.35	20
Sub-base	104	0.35	25
Sub-grade	34.5	0.45	200

2.5.1.1 Effect of Vehicle Wheel Position

Crack tip parameters including mode I stress intensity factor (i.e. K_I), mode II stress intensity factor (i.e. K_{II}), and T-stress (i.e. non-singular stress) are given in Fig. 2.18. According to Fig. 2.18a, K_I had a negligible value as the vehicle wheel located far from the crack. As the vehicle wheel moved towards the crack, its value firstly increased and then, before locating upon the crack (i.e. $L = 0$), decreased to zero. Therefore, the crack faces were completely closed (i.e. $K_I = 0$) as the vehicle wheel located close to the crack. Distribution of stresses inside the HMA layer loaded by a

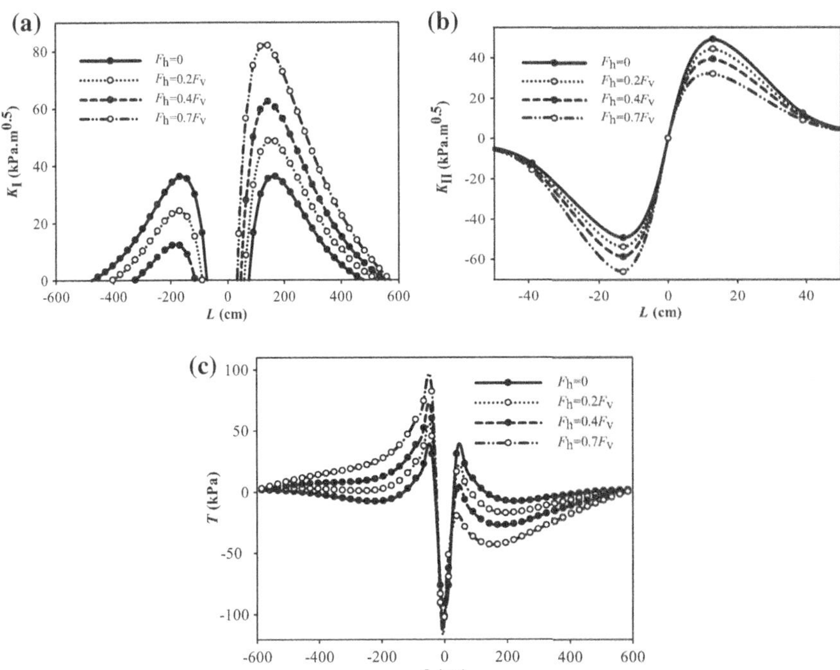

Fig. 2.18 Values of the crack tip parameters including **a** K_I, **b** K_{II}, **c** T-stress versus L for different horizontal loads [42]

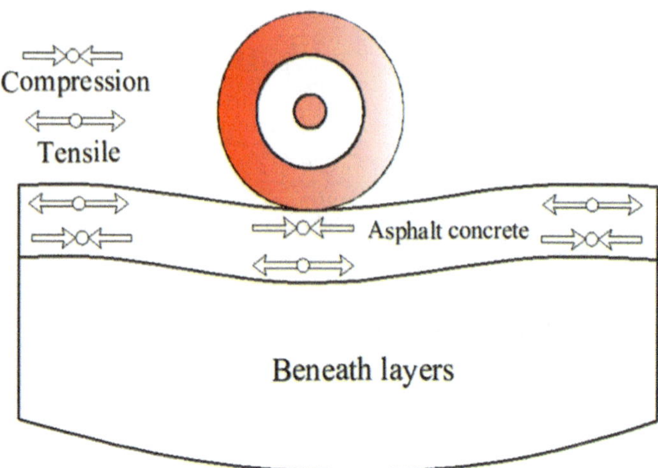

Fig. 2.19 Stress distribution within the HMA layer loaded by a vehicle wheel [42]

vehicle wheel confirmed this result. According to Fig. 2.19, the compressive stresses under the vehicle wheel closed the crack faces (i.e. $K_I = 0$). While, the vehicle wheel produced tensile stresses at other regions near the road surface, leading to opening the crack faces (i.e. $K_I > 0$). It is also pointed out that the values of K_I were identical at the right and left sides of the crack plane because of the symmetry.

As a vehicle passed on a road, the shear deformation mode (i.e. K_{II}) was also observed at the crack tip. The results of K_{II} are given in Fig. 2.18b. Based on this figure, the shear deformation mode occurred only where the vehicle wheel was very close to the crack; while, the crack experienced no opening deformation mode (i.e. $K_I = 0$) at this region. Consequently, the regions around a crack present in asphalt pavement may be classified into three zones:

(i)   First zone: which corresponds to the regions far enough from the crack, and only pure mode I deformation was observed at the crack tip.
(ii) Second zone: which corresponds to the regions very close to the crack, and only pure mode II deformation was observed at the crack tip.
(iii) Third zone: which corresponds to the regions in between, and mixed I/II deformation mode was observed at the crack tip.

As the vehicle wheel approached the crack, the value of K_{II} (with negative sign) initially increased, and then decreased to zero where the wheel was positioned exactly upon the crack. On the other hand, as the wheel moved far away from the crack, the value of K_{II} again increased, and then decreased, while its sign reversed and was switched to positive. The sign change in the value of K_{II} signifies that the direction of the shear deformation has been changed; consequently, the passage of vehicles on the cracked road can lead to the fatigue crack growth.

2.5.1.2 Effect of Horizontal Load

The effect of horizontal load on the crack tip parameters was also investigated in [42]. The horizontal load F_h has a value between 0 and 70% of vertical load F_h depending on the severity of acceleration and deceleration of a vehicle [40]. Accordingly, the horizontal load F_h was varied as 0, $0.2F_v$, $0.4F_v$ and $0.7F_v$; while, the vertical load was considered constant as $F_v = 40$ kN. The values of K_I and K_{II} resulting from passing a vehicle wheel on a road are displayed in Fig. 2.18, in which the trend of K_I variations for the cases that both the vertical and horizontal loads were simultaneously applied on the road was similar to the case that the vertical load was solely applied on the road (i.e. the case of $F_h = 0$). In other words, as the vehicle wheel moved towards the crack plane, K_I initially increased and then decreased to zero. On the other hand, the value of K_I for the cases that both the horizontal and vertical loads were simultaneously applied on the road was smaller than that for the case of applying the vertical load solely. When the vehicle wheel was located on the right side of the crack, the results were reversed i.e. enhancement of the horizontal load increased the value of K_I significantly, such that the horizontal load equaling the values of $0.2F_v$, $0.4F_v$ and $0.7F_v$ increased the peak value of K_I by 34%, 72%, and 125%, respectively. On the other hand, the horizontal load made the crack faces to open at lower values of L, and to close at higher values of L (see Fig. 2.18a). Thus, higher values of horizontal load caused the crack faces to open at wider range of L.

Likewise, the trend of K_{II} variations for the cases that both the horizontal and vertical loads were simultaneously applied on the road was similar to the case of applying the vertical load solely, which was explained above. For the vehicle wheel positions of $L < 0$, the presence of horizontal load increased the value of K_{II}; whereas, for the cases of $L > 0$, the value of K_{II} decreased. The horizontal load with the values of $0.2F_v$, $0.4F_v$ and $0.7F_v$ increased the peak value of K_{II} by 10%, 20%, and 35%, respectively. From Fig. 2.18, it is also seen that the horizontal load has influenced K_I more than K_{II}.

In addition to the mode I and mode II SIFs (i.e. K_I and K_{II}), the value of T-stress was also obtained from the finite element analyses. K_I and K_{II}, describing the stress singularity surrounding the crack tip, control the onset of crack growth. Furthermore, T-stress and its sign influence the onset of mixed-mode I/II fracture in the cracked components significantly. It is noticed that the negative value of T-stress increases the load bearing capacity of cracked parts; while, the positive T-stress decreases the load bearing capacity of cracked parts subjected to mixed-mode I/II loading. Figure 2.18c exhibits the results of T-stress for different values of L and F_h. Similar to the results of K_I and K_{II}, T-stress had a value of zero when the wheel located far enough from the crack. As the wheel approached the crack, T-stress initially increased and then decreased sharply. Furthermore, for the case of $L < 0$, T-stress increased remarkably as the horizontal load enhanced; while, its manner reversed for the wheel positions of $L > 0$. From the above explanations, it can be concluded that T-stress at the crack tip can accelerate the crack propagation in the asphalt pavements.

The horizontal load resulting from vehicle acceleration or deceleration increased the crack tip parameters of K_I, K_{II} and T-stress significantly. Hence, a cracked asphalt concrete is subjected to critical loading condition where vehicles brake or accelerate (for example, at the road junctions). Consequently, an asphalt overlay with higher resistance to fracture should be used in these areas.

2.5.1.3 Effect of Elasticity Modulus

Elasticity modulus of road layers is another important parameter influencing the crack tip parameters (i.e. SIFs and T-stress). The effect of elasticity modulus of all pavement layers illustrated in Fig. 2.17 was evaluated at three levels of E°, $5E^\circ$ and $10E^\circ$ in [42]. E° refers to the reference elasticity modulus of road layers given in Table 2.3. This range of elasticity modulus (E) considered in this research arises from a fact that temperature or aging changes the elasticity modulus of asphalt mixtures significantly. For example, by decreasing the ambient temperature from 25 to $-$ 10 °C, the elasticity modulus of asphalt concrete increases over 10 times [47]. The elasticity modulus of beneath layers of asphalt pavement can also be changed by the use of materials other than asphalt concrete such as cement concrete. It is pointed out that for all the investigations performed herein, the elasticity modulus of the studied layer was varied as mentioned above (i.e. E°, $5E^\circ$ and $10E^\circ$), while that of other layers was regarded as presented in Table 2.3 (i.e. E°). Besides, for each elasticity modulus, the analyses were performed at three levels of the horizontal loads (i.e. 0, $0.4F_v$ and $0.7F_v$) along with the vertical load.

Elasticity Modulus of Asphalt Concrete

Figure 2.20 displays the value of K_I for different elasticity moduli of the asphalt concrete at different locations of the vehicle wheel (i.e. L). For the case of $F_h = 0$, as the elasticity modulus E enhanced, the mode I stress intensity factor K_I increased, and the increase of K_I took place at smaller value of L. The peak value of K_I for the cases of $5E^\circ$ and $10E^\circ$ with respect to that for the case of E° was 3 and 5, respectively. Thus, the value of K_I depended highly on the elasticity modulus of asphalt concrete; hence, decrease in the ambient temperature can expose the asphalt concrete layer in a critical condition of crack extension.

According to Fig. 2.20a, the value of K_I for the higher values of asphalt concrete elasticity modulus was less influenced by the horizontal load than for the lower values of asphalt concrete elasticity modulus. For example, the ratio of $\dfrac{K_I^{\max} \text{ (for } F_h = 0.7F_v)}{K_I^{\max} \text{ (for } F_h = 0)}$ for the asphalt concrete with the elasticity modulus of E°, $5E^\circ$ and $10E^\circ$ was found as 2.26, 1.57 and 1.38, respectively.

The results of K_{II} for different elasticity moduli of the asphalt concrete at different locations of the vehicle wheel (i.e. L) are given in Fig. 2.20b, in which enhancement of the asphalt concrete elasticity modulus increased the value of K_{II}. For example, the peak value of K_{II} (at the wheel locations of $L < 0$) for the cases of $5E^\circ$ and $10E^\circ$ with respect to that for the case of E° increased by 38% and 51%, respectively. The

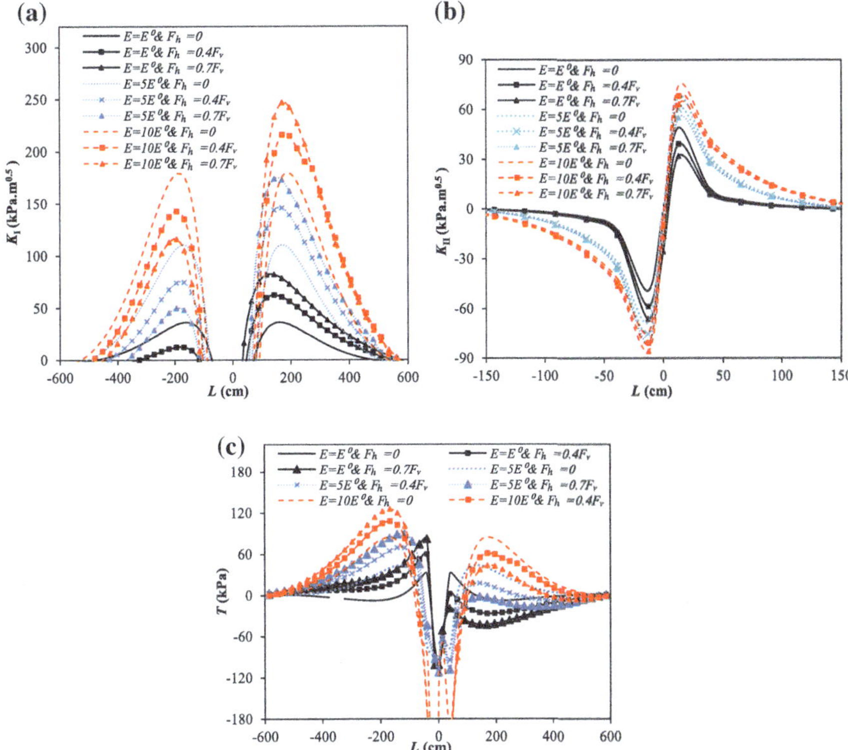

Fig. 2.20 Crack tip parameters of **a** K_I, **b** K_II, **c** T-stress versus L for different asphalt concrete elasticity modulus and horizontal loads [42]

presence of horizontal load increased the value of K_II for all the cases of E°, $5E^\circ$ and $10E^\circ$. For example, the ratio of $\dfrac{K_\mathrm{II}^{\max}{}_{(\text{for } F_\mathrm{h}=0.7F_\mathrm{v})}}{K_\mathrm{II}^{\max}{}_{(\text{for } F_\mathrm{h}=0)}}$ for the cases of E°, $5E^\circ$ and $10E^\circ$ was found as 1.35, 1.19 and 1.15, respectively. Hence, similar to the results observed for K_I, the value of K_II for the higher values of asphalt concrete elasticity modulus was less influenced by the horizontal load than for the lower values of asphalt concrete elasticity modulus.

The results given above indicates that both the horizontal load and elasticity modulus of asphalt concrete influenced the values of K_I more than the values of K_II.

The results of T-stress for different elasticity moduli of asphalt concrete at different locations of the vehicle wheel (i.e. L) are also given in Fig. 2.20c, in which T-stress was highly influenced by both the asphalt concrete elasticity modulus and the horizontal load.

Elasticity Modulus of Base Layer

In order to evaluate the effect of elasticity modulus of base layer on the crack tip parameters, the elasticity modulus of base layer was varied at three levels of $E°$, $5E°$ and $10E°$ in the finite element analyses while the mechanical properties and dimensions of all other road layers were those given in Table 2.3. Furthermore, the analyses were carried out at three different levels of horizontal load (i.e. $F_h = 0$, $0.4F_v$ and $0.7F_v$) for each level of the base layer elasticity modulus.

The results for the case of $F_h = 0$ exhibited that the base layer elasticity modulus affected the parameter of K_I slightly. For the cases of $F_h = 0.4F_v$ and $F_h = 0.7F_v$ and where the wheel located at $L > 0$, increase in the base layer elasticity modulus resulted in the reduction of K_I. This is in contrast to the results observed for the asphalt concrete layer as shown in Fig. 2.20a.

Finite element results also exhibited that the increase of base layer elasticity modulus led to decrease in the value of K_{II} for all the cases of horizontal load. Considering the results of K_I and K_{II} indicated that the destructive effect of horizontal load at junctions, bus stations, etc. can be declined by the use of materials with higher elasticity modulus (for example, cement concrete instead of asphalt concrete) in the base layer of a road. While, the use of a material with higher elasticity modulus in the top layer of a road produced higher SIFs at the crack tip (see Fig. 2.20).

Unlike the results observed for the asphalt concrete, increase of the base layer elasticity modulus led to reduction of T-stress for all the horizontal loads. Meanwhile, the peak value of T-stress took place at the same value of L and very close to the crack plane.

Elasticity Modulus of Sub-base Layer

The values of crack tip parameters for different elasticity moduli of the sub-base layer and horizontal loads at different locations of the vehicle wheel (i.e. L) were also obtained in [42]. According to the results of K_I for the case of $F_h = 0$, increase of the sub-base elasticity modulus from $E°$ to $5E°$ resulted in the reduction of peak value of K_I, and it nearly remained constant as the sub-base elasticity modulus was varied from $5E°$ to $10E°$. It is also noticed that increase of the sub-base elasticity modulus led to decrease of K_I at the corresponding horizontal load, similar to the results observed for the base layer.

As such, increase in the elasticity modulus of sub-base layer resulted in the reduction of K_{II} at the corresponding horizontal load, similar to the results observed for the base layer. Hence, stress intensity factors (i.e. K_I and K_{II}) at the crack tip can be decreased through the increase of elasticity modulus of sub-base layer, particularly when both of the vertical and horizontal loads are applied to a road surface.

When the wheel was located at $L < -120$ cm, increase in the sub-base elasticity modulus led to decrease of T-stress at the corresponding horizontal load. This behavior reversed at $L = -120$ cm, and therefore T-stress increased as the elasticity modulus enhanced.

Elasticity Modulus of Sub-grade Layer

On the basis of results given in [42], increase of the sub-grade elasticity modulus led to decrease of K_{I}; while, the parameter K_{II} was not influenced by varying the elasticity modulus of the sub-grade layer. Furthermore, T-stress was not approximately affected by the increase of the sub-grade elasticity modulus, and it was only influenced by the horizontal load i.e. enhancement of the horizontal load resulted in the increase of T-stress.

By comparing the results of SIFs, it is clear that further reduction in the values of K_{I} was achieved by increasing the elasticity modulus of beneath layers. In addition, as increase in the elasticity modulus took place at lower layers, its effect on K_{II} and T-stress diminished. For example, change in the elasticity modulus of the sub-grade layer did not influence the values of K_{II} and T-stress.

From the explanations mentioned above, it is concluded that in order to avoid the crack growth at junctions, bus stations, etc., a material having low elasticity modulus should be used at the top layer of a road (for example, the use of modified asphalt mixtures), and a material having high elasticity modulus should be used at beneath layers including base, sub-base and sub-grade layers.

2.5.1.4 Effect of Thickness of Road Layers

The effect of thickness of pavement layers (i.e. asphalt concrete, base and sub-base) on the crack tip parameters have been investigated in [42]. The thickness of the layers was assessed at three levels such that the thickness of the studied layer would be given below, and the thickness of the other layers was considered as those given in Table 2.3. Besides, for each thickness of pavement layer, the investigations were performed at three levels of the horizontal loads (i.e. 0, $0.4F_{\mathrm{v}}$ and $0.7F_{\mathrm{v}}$) along with the vertical load.

In order to evaluate the effect of the thickness of asphalt concrete layer on the crack tip parameters, it was varied as 14, 20 and 27 cm. Figure 2.21 plots the results of the crack tip parameters, in which by increasing the thickness of the asphalt concrete layer, all the crack tip parameters including K_{I}, K_{II} and T-stress decreased. In addition, increase in the horizontal load enhanced the crack tip parameters significantly for some of the vehicle wheel positions.

In the next step, thickness of the base layer was varied as 20, 27 and 34 cm. Increase of the thickness of the base layer resulted in the reduction of K_{I} for all the horizontal loads. On the basis of finite element results, K_{II} and T-stress were not sensitive to change in the base layer thickness, and only depended on the horizontal load. In other words, enhancement of the horizontal load resulted in the increase of K_{II} and T-stress.

The effect of the thickness of the sub-base layer was finally assessed in [42]. It was varied as 25, 35 and 45 cm in the finite element analyses. The results indicated that K_{I} reduced as the sub-base thickness increased. Meanwhile, K_{II} and T-stress were not affected by changing the base layer thickness, and only depended on the

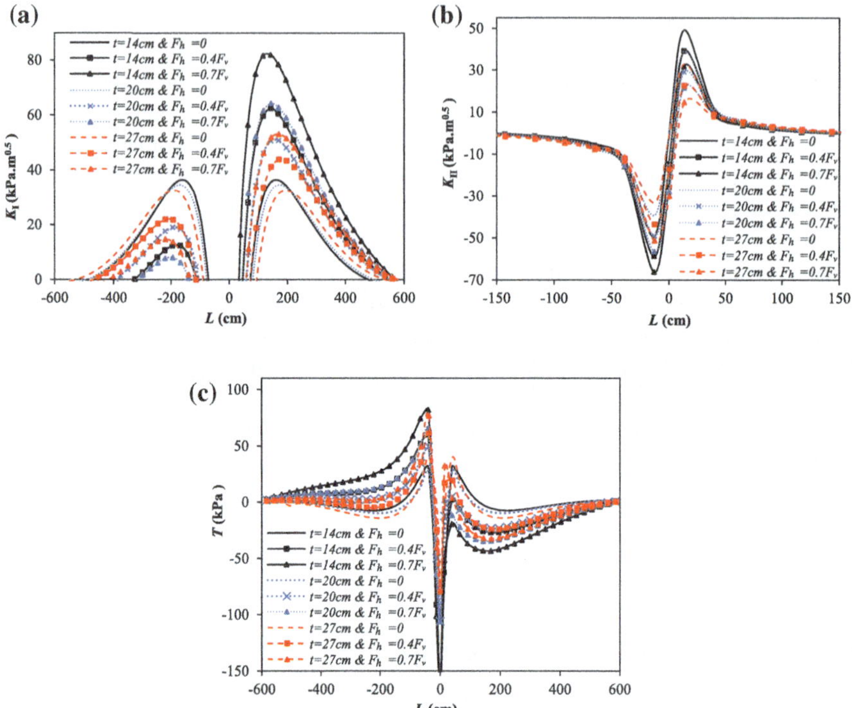

Fig. 2.21 Crack tip parameters of **a** K_I, **b** K_{II}, **c** T-stress versus L for different asphalt concrete thicknesses and horizontal loads [42]

horizontal load. In other words, enhancement of the horizontal load resulted in the increase of K_{II} and T-stress.

To summarize the effect of the thickness of road layers, it can be said that the crack tip parameters reduced as the thickness of road layers increased. In general, the values of K_I and K_{II} were more sensitive to the thickness of the asphalt concrete layer than that of the base and sub-base layers.

It is also worth mentioning that the values of crack tip parameters obtained from the finite element analyses are less than the fracture toughness of asphalt concretes, which its value has been reported between 0.6 and 1.2 MPa $m^{0.5}$ as per the previous investigations [48–52]. Hence, the pass of only one vehicle cannot lead to the sudden fracture in road structure; however, repetitive traffic loads may result in the fatigue crack growth.

In another investigation performed by Aliha and Sarbijan [53], the effect of crack type (i.e. top-down and bottom-up crack) has been studied. For the comparison purpose, they assumed that the lengths and locations of both cracks were the same. According to their results, the values of K_I and T-stress for the top-down crack were higher than the corresponding values for the bottom-up crack. In other words, mode

I stress intensity factor and T-stress were more sensitive to the top-down crack than to the bottom-up one.

Conversely, the effect of mode II component (i.e. K_{II}) was more pronounced where a bottom-up crack existed in the pavement. They finally concluded that a top-down crack was more vulnerable to propagation through the thickness of pavement under repetitive vehicular loads in comparison to a bottom-up crack. It is also pointed out that the similar results for the effect of bottom-up crack on the crack tip parameters has been reported by Rostami [54].

2.5.2 Three-Dimensional Investigations

Two-dimensional finite element modeling of road structures does not regard significant aspects of crack growth such as mode III deformations. Recent developments on the finite element modeling of road structures indicate that the 3D finite element analysis offers more accurate and reliable results as compared with the 2D finite element analysis.

Ayatollahi et al. [55] determined crack parameters at crack front of a road structure with four layers of asphalt concrete, base, sub-base, and sub-grade by performing 3D finite element analyses. In this study, the top layer, namely asphalt concrete, contained an edge top-down crack of length $c = 130$ cm and depth $a = 12$ cm. The crack configuration and the location of wheels are illustrated in Fig. 2.22. The geometry parameters shown in Fig. 2.22 are able to consider the main features of the studied crack.

As illustrated in Fig. 2.22, a vehicular loading with 80 kN of single-axle was regarded for the finite element analyses, while each tyre applied 40 kN to the surface of road. It is pointed out that the mentioned load is a typical value applied by trucks on the road surface. The contact of tyre on the road was assumed as rectangular area of 18 cm $\times$ 26 cm, and each wheel applied a pressure of 855 kPa. The distance between the right and the left wheels was assumed 200 cm, which is a typical value in trucks. The location of the wheels was described by parameters D and L, which respectively correspond to the right wheel distance from the edge of road and the distance between the wheels axle and the crack plane (see Fig. 2.22a). The crack was assumed that has been started from the road edge within the layer of asphalt concrete and was propagating towards another edge of the road. Hence, the crack head (i.e. the quarter-circle shown in Fig. 2.22b) was the focal zone which the crack parameters were computed. The radius r of the crack head was assumed to be 12 cm. The location of a point on the crack head was described by angle φ, so the values of $\varphi = 0°$ and $\varphi = 90°$ correspond to the points at the road surface and the deepest point, respectively. The length and the width of the studied road structure, l and w, were respectively assumed to be 900 and 600 cm. It is pointed out that the tyre location can be controlled by changing the values of D and L.

As mentioned earlier, crack growth can take place under different deformation modes including mode I, mode II, mode III or combinations of them. A 3D finite

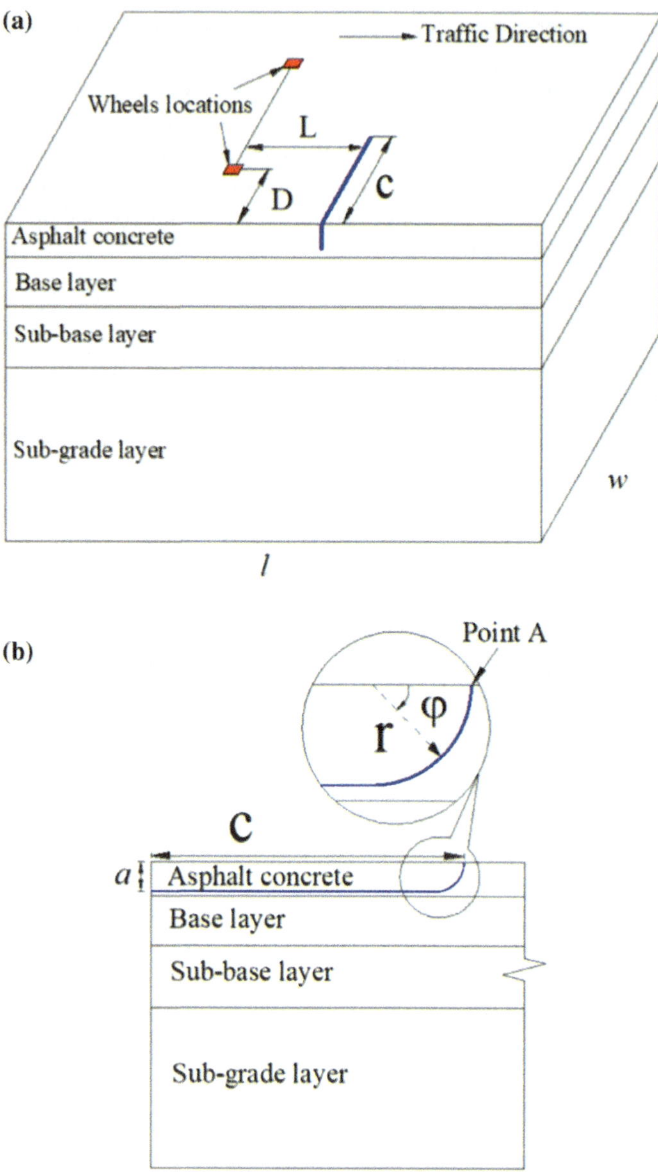

Fig. 2.22 **a** Cracked road structure, **b** side view of a crack within the asphalt concrete [46]

element model was described herein in order to study the contributions of these modes in development of an edge crack within a road structure.

All the road layers including asphalt concrete, base, sub-base and sub-grade were assumed as homogenous, isotropic and linear elastic in the macro-scale. These assumptions have been frequently used by other scholars (see e.g. [43–45]). The mechanical properties and the thickness of the road layers studied in [55] are presented in Table 2.3.

In the finite element simulations, the boundary conditions were considered such that the side faces (i.e. front, rear, left and right faces) were fixed normal to the cut faces, while the other two degrees of freedom were free. As such, the bottom face (i.e. the bottom of the sub-grade layer) was completely fixed in all directions.

The finite element code ABAQUS was used for three-dimensional models of the road structure. These models were generated by using the second-order elements (i.e. 20-node bricks). The stress intensity factors K_I, K_{II} and K_{III} were directly extracted from ABAQUS which uses an interaction integral method to calculate the stress intensity factors for a crack subjected to mixed-mode loading.

Singular elements with nodes at quarter-point positions were employed for the first ring of elements around the crack front. The contour integrals were obtained in the circular zone surrounding the crack front. No significant variation in the contour integral values was observed according to the results of finite element analyses. A mesh-convergence analysis was also conducted to obtain the suitable size and number of elements for constructing finite element models.

A large number of 3D finite element simulations were carried out on the cracked road structure shown in Fig. 2.22a. The stress intensity factors K_I, K_{II} and K_{III} were obtained at the points of the crack head displayed in Fig. 2.22b (i.e. along the quarter-circular arc of the crack front). These points were selected for stress intensity factor calculations because the crack was expected to extend towards the centerline of road. It is reminded that L represents the distance between the wheels axis and the crack plane. For the case of $L = 0$ and $D = 0.2$ m, the wheels were symmetric relative to the point A shown in Fig. 2.22. This symmetry disappeared as the value of D increased, and the right wheel located upon the point A for the case of $D = 1$ m. The finite element results, in terms of stress intensity factors, computed along the crack head for different wheel locations (i.e. L and D) are given in the following sections.

2.5.2.1 Mode I Stress Intensity Factor

Figure 2.23 plots the values of mode I stress intensity factor K_I along the crack head (i.e. the angle φ) for different values of L and D. Based on Fig. 2.23a, for the case of $D = 0.2$ m, as the wheels approached the top-down crack, the value of K_I enhanced significantly. When the distance of wheels was far from the crack (i.e. when $L > 3.5$ m), the value of K_I was nearly zero. Hence, the crack nearly experienced no mode I deformation when it located far enough from the wheels. In addition, for any value of L, K_I exhibited a maximum value at a certain value of φ. As the value of φ increased, K_I firstly enhanced to reach its maximum value, and then reduced.

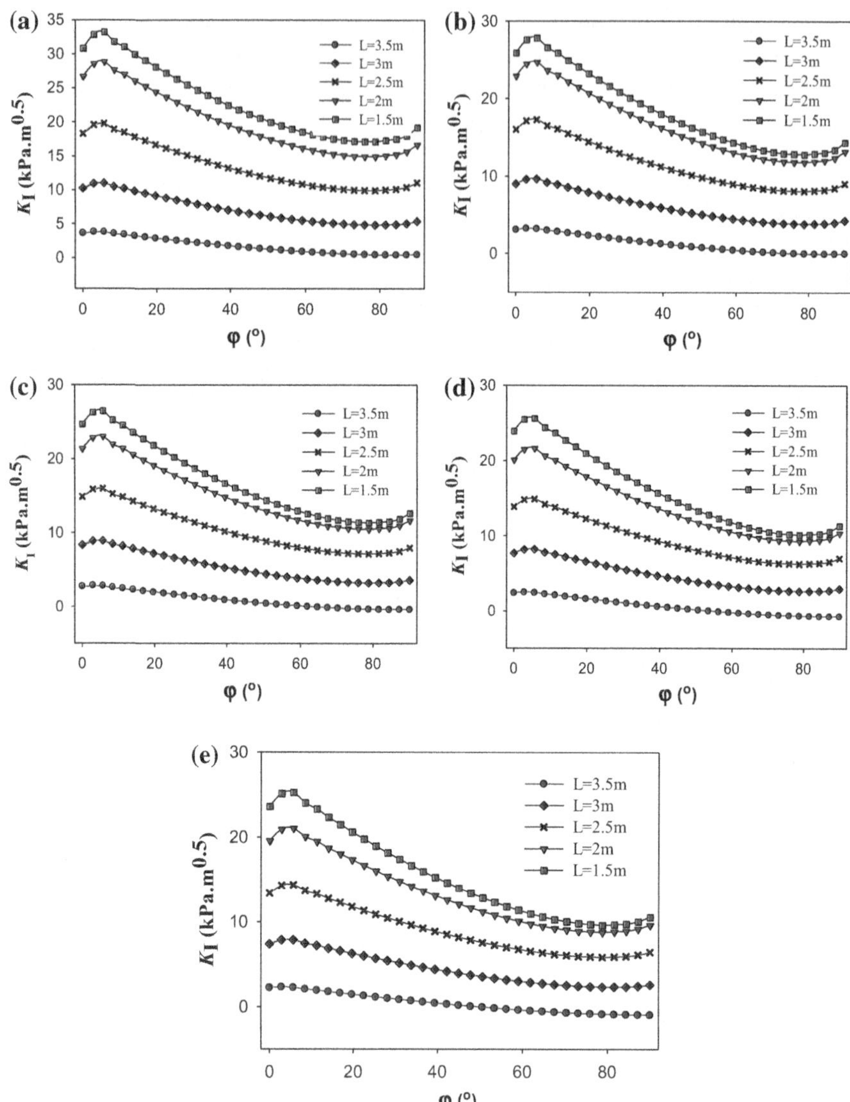

Fig. 2.23 Mode I stress intensity factor for different values of φ, L and **a** $D = 0.2$ m, **b** $D = 0.5$ m, **c** $D = 0.7$ m, **d** $D = 0.9$ m, **e** $D = 1$ m [46]

Meanwhile, by moving the wheels towards the crack plane, the slope of K_I variations against φ intensified, and the maximum K_I took place at about $\varphi = 5.6°$ for all values of L.

For the cases that the wheels were near the crack plane (i.e. when $L < 1.5$ m), the top-down crack closed due to downward deformations occurring in asphalt concrete layer, which resulted in negative values of K_I. This result is identical to that observed by Ameri et al. [41]. However, K_I was assumed to be zero for a closed crack, since its effect on crack growth was negligible.

The mode I stress intensity factor K_I generally reduced by increasing the value of φ. This signifies that the critical condition along the crack head occurred near the road surface, and therefore the crack tended to develop transversely towards the centerline of the road rather than developing vertically towards the beneath layers.

Finite element results for the cases of $D = 0.5, 0.7, 0.9$ and 1 m are respectively shown in Fig. 2.23b–e. The results of K_I for the mentioned values of D were identical to those for the case of $D = 0.2$ m. Meanwhile, the value of K_I reduced by increasing the value of D for each fixed value of L.

According to the abovementioned results, the opening deformation mode was observed for a wide location of the wheels on the road (i.e. wide range of L and D); hence, this mode can be a major threat for the crack propagation and the deterioration of road structures. Deterioration resulting from cracking is in particular of great importance for cold climate regions in which asphalt mixtures are less deformable and so brittle fracture and overall failure becomes more probable. The investigations performed in recent decades have been often focused on the mode I fracture of asphalt concretes, and extensive efforts have been also made to improve fracture resistance of asphalt concretes for reducing their damage against the mode I fracture (see e.g. [56, 57]). Indeed, asphalt concretes are generally composed of aggregate, air void, binder, and modifier, and their fracture behavior is highly influenced by these ingredients. Hence, the fracture resistance of asphalt concretes can be significantly improved by selecting proper combinations of these ingredients as well as by employing suitable binder modifiers (see e.g. [50, 58–61]).

2.5.2.2   Mode II Stress Intensity Factor

Finite element results given in [55] indicated that in addition to the opening deformation mode (i.e. mode I), the in-plane shear deformation mode (i.e. mode II) occurred at the crack front as well. Figure 2.24 presents the values of mode II stress intensity factor K_{II} along the crack head (i.e. the angle φ) for different values of L and D. It is reminded that for the shear deformation modes, both positive and negative signs of K_{II} and K_{III} contribute in the process of crack extension, and the only difference is that the relative directions of the crack face deformation reverse by switching the sign of K_{II} or K_{III}. However, for deformation mode of I, the positive and negative signs correspond to opening and closing the crack faces, respectively, and only opening the crack faces (i.e. the positive sign of K_I) contributes in the process of crack propagation.

According to Fig. 2.24a, for the case of $D = 0.2$ m, as the wheels approached the top-down crack, the value of K_{II} firstly enhanced; however, the value of K_{II} reduced for lower values of L. Similar to the results observed for the opening deformation mode, the effect of traffic load on the in-plane shear deformation mode was negligible

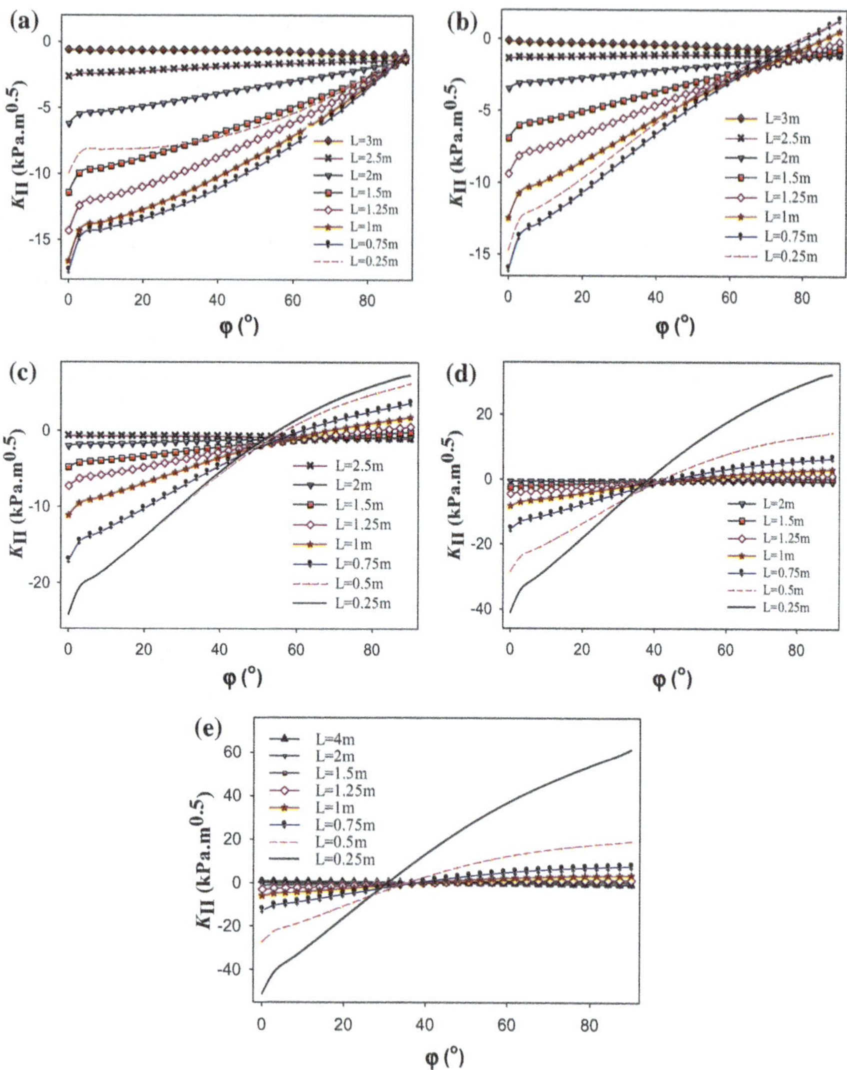

Fig. 2.24 Mode II stress intensity factor for different values of φ, L and **a** $D = 0.2$ m, **b** $D = 0.5$ m, **c** $D = 0.7$ m, **d** $D = 0.9$ m, **e** $D = 1$ m [46]

for the cases that the wheels located far enough from the crack plane (i.e. for $L >$ 3 m). Furthermore, as the value of φ increased, the magnitude of K_{II} reduced sharply, and then decreased gradually to reach nearly zero at $\varphi = 90°$ for all the values of L. Similar to the results observed for the mode I deformations, the values of K_{II} in the regions near the road surface were higher than those in the deepest regions of the crack head (i.e. higher values of φ). As a result, combination of the mode I and mode II stress intensity factors can make the top-down cracks to propagate towards the centerline of roads in the transverse direction rather than propagating downwards. Other researchers have reported similar observations; for example, Buttlar et al. [62] indicated that a crack propagated in the transverse direction through their field investigations.

Figure 2.24b plots the results of K_{II} for the case of $D = 0.5$ m. These results generally were similar to the case of $D = 0.2$ m, with this important difference that K_{II} decreased to zero at the angles about $\varphi = 75°$, and then started to increase while its sign changed from negative to positive. Hence, the direction of in-plane shear deformations did not remain constant for the whole values of φ i.e. it changed at a specific value of φ depending on the wheels locations (namely L). It is concluded from comparing Fig. 2.24a–e that as the value of D increased, the sign of K_{II} changed at lower values of φ. Meanwhile, the maximum value of K_{II} (which was nearly 50 kPa m$^{0.5}$) occurred at a wheels position of $D = 1$ m and $L = 0.25$ m i.e. when one of the wheels located very close to the crack head. According to Fig. 2.23, the maximum value of K_I was about 33 kPa m$^{0.5}$. Therefore, a crack within the road may experience higher mode II deformation than mode I deformation for some particular wheel locations. Based on the results presented in [55], as a vehicle passed over a cracked road, the crack experienced mixed mode I/II deformation. In this regard, many experimental studies have been performed on the mode I (in the past decades, see e.g. [63–66]) and mixed mode I/II loadings (in recent years, see e.g. [48, 49, 51, 67]) on asphalt concretes.

2.5.2.3   Mode III Stress Intensity Factor

According to 3D finite element analyses performed by Ayatollahi et al. [55], as a vehicle passed on a cracked road, in addition to opening (i.e. K_I) and in-plane shear (i.e. K_{II}) deformation modes, the out-of-plane shear deformation mode (i.e. K_{III}) was also observed along the crack front. Figure 2.25a exhibits the values of mode III stress intensity factor K_{III} through the crack head (i.e. the angle φ) for different values of L and D. Similar to the results of K_I (shown in Fig. 2.23a) and K_{II} (shown in Fig. 2.24a), the mode III stress intensity factor K_{III} was also negligible when wheels located far enough from the crack plane (i.e. when $L > 3$ m). As the wheels approached the crack plane, the value of K_{III} increased for all values of φ (except for $L = 0.25$ m). Generally, by increasing φ, the value of K_{III} initially reduced, and then increased. As is evident from Fig. 2.25a, the maximum K_{III} was achieved at $\varphi = 90°$ for the wheels location of $D = 0.2$ m. The results of K_{III} for the wheels position of $D = 0.5$ m are shown in Fig. 2.25b. The results were identical to the case of $D = 0.2$ m i.e. by increasing φ, the value of K_{III} initially reduced, and then enhanced gently; however, the maximum K_{III} occurred at $\varphi = 0°$ (namely, on the road surface). The general trends for the results of K_{III} in the cases of $D = 0.7, 0.9,$ and 1 m were different from those for the cases of $D = 0.2$ and 0.5 m. However, the maximum K_{III} was still achieved at the road surface (i.e. at $\varphi = 0°$). Based on the results, a change in the sign of K_{III} was observed for some positions of wheels.

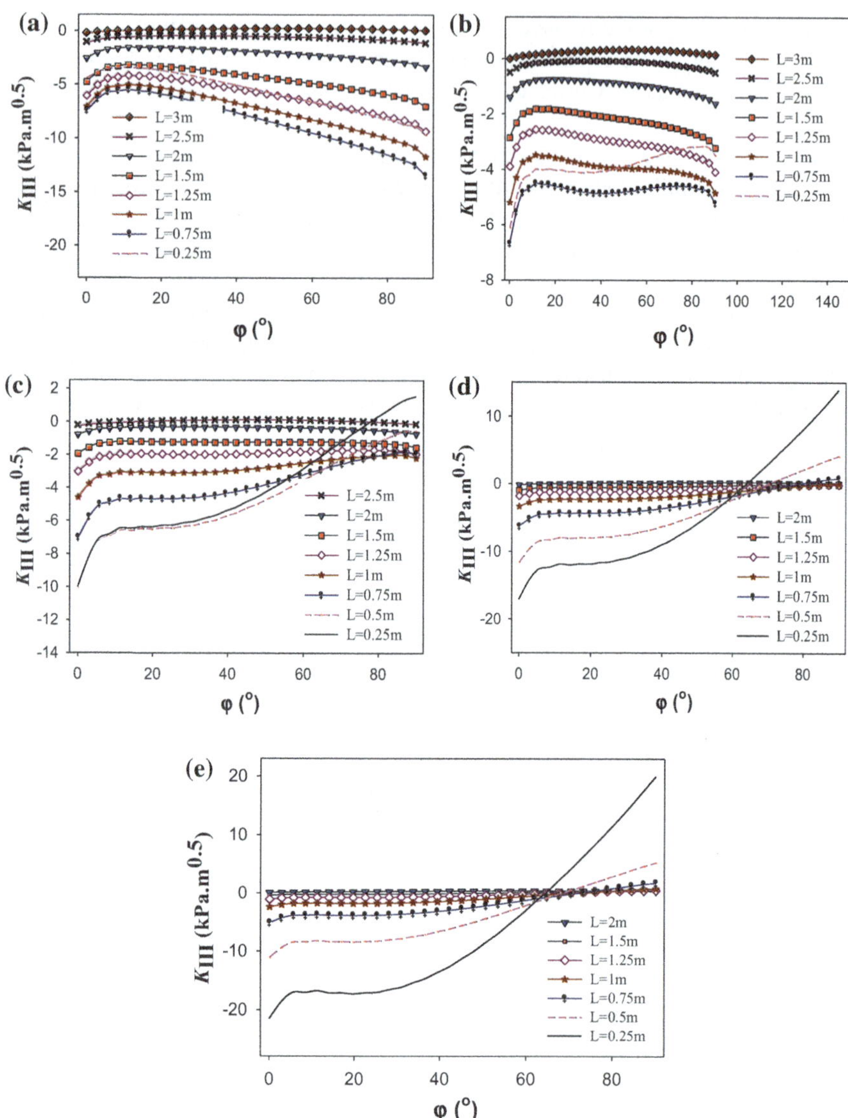

Fig. 2.25 Mode III stress intensity factor for different values of φ, L and **a** $D = 0.2$ m, **b** $D = 0.5$ m, **c** $D = 0.7$ m, **d** $D = 0.9$ m, **e** $D = 1$ m [46]

In addition, as a vehicle got away from the crack plane (i.e. for higher values of D), the value of φ at which the sign of K_{III} reversed, reduced.

Ayatollahi et al. [55] reported that for all values of D, by approaching a vehicle to the crack plane (i.e. when L reduced), the value of K_I initially enhanced, and then reduced until the crack faces closed completely. For the wheels position of $|L|$

< 0.5 m, the sign of K_{I} was negative, signifying that the mode I deformation had no effect on the crack propagation. A similar trend was seen for the mode II stress intensity factor K_{II}. In other words, by decreasing the longitudinal distance of wheels (i.e. when L reduced) relative to the crack plane, the value of K_{II} initially enhanced, and then sharply reduced to zero for all values of D. Furthermore, the results of K_{III} were very similar to the results of K_{II}, but the values of K_{III} were relatively less than those of K_{II}. The loads inserted to the crack were symmetric when a vehicle located exactly upon the crack line (i.e. when $L = 0$). Hence, in this loading condition, no relative sliding or tearing deformations occurred (i.e. $K_{\mathrm{II}} = K_{\mathrm{III}} = 0$) between the crack faces, as expected.

Finite element analyses demonstrated that the vehicle position on the road resulted in different deformation modes of I, II and III at the crack front. Ayatollahi et al. [55] also investigated the portion of shear deformations including in-plane and out of plane modes in the crack extension. For this purpose, the contribution of shear deformation (i.e. the ratio of shear deformation to the total deformation) was calculated through the energy release rate G, which can be defined as follows:

$$\frac{G_{\mathrm{II}} + G_{\mathrm{III}}}{G} \tag{2.18}$$

where, $G_{\mathrm{I}} = \frac{K_{\mathrm{I}}^2}{E}$, $G_{\mathrm{II}} = \frac{K_{\mathrm{II}}^2}{E}$, $G_{\mathrm{III}} = \frac{K_{\mathrm{III}}^2}{2\mu}$ and $G = G_{\mathrm{I}} + G_{\mathrm{II}} + G_{\mathrm{III}}$. Also, E and μ are the Young's modulus and the shear modulus, respectively.

According to the 3D finite element results given in [55], when the wheels located far from the crack plane (i.e. high values of L), shear deformations were negligible compared to the total deformations at crack front. As the wheels approached the crack plane, the shear deformations became dominant relative to the opening deformation, and when the location of wheels was very close to the crack plane, the total deformation was nearly related to the shear modes. As a result, the shear deformation modes can play an important role in propagation of top-down crack under traffic loading. Particularly, the mode III stress intensity factor K_{III} can significantly affect the process of crack propagation in the transverse direction.

In another investigation, Ameri et al. [41] performed a 3D finite element analysis on a cracked road structure having four layers of asphalt concrete, base, sub-base and sub-grade. They considered a top-down transverse crack within the asphalt concrete layer in the middle of the road, as shown in Fig. 2.26. Four wheels, representing the total loads applied from a vehicle to the road surface, were assumed to insert loads on the road surface. The wheels location relative to the crack plane was defined by parameters D and L. Based on the results of this investigation, the crack experienced different deformation modes of I, II and III. When a vehicle moved towards the crack plane, the mode I stress intensity factor K_{I} firstly increased, and as the wheels approached further to the crack plane, the K_{I} dramatically reduced and its sign changed from positive to negative. In addition, while the mode I deformation was still high for greater values of L, the shear deformations (i.e. mode II and mode III SIFs) became significant only when the wheels were close to the crack plan. Generally, the

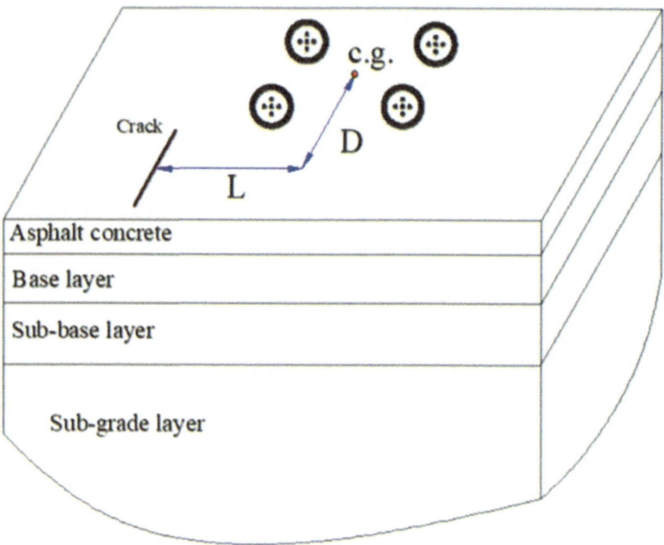

Fig. 2.26 Cracked road structure subjected to loading by a four-wheel vehicle [41]

effect of opening deformation mode was more pronounced than the shear deformation modes, although the magnitudes of K_{II} and K_{III} were still significant. These trends for the SIFs are similar to those observed by Ayatollahi et al. [55], as discussed earlier.

2.6 Extended Finite Element Method (XFEM) in Modeling Fracture of Asphalt Concretes

The extended finite element method (XFEM), also known as generalized finite element method is a numerical technique extending the classical finite element method by developing the solution space to differential equations with discontinuous functions. The XFEM has been extended to simplify difficulties in solving problems with localized features (such as interfaces, voids, cracks, etc.) that are not efficiently resolved by mesh refinement. One of the main applications of the XFEM is to model fracture of materials, in which discontinuous fundamental functions are combined with standard polynomial functions for elements that are intersected by a crack. A key advantage of XFEM is that in such problems the finite element mesh does not require to be updated to track the crack path. It has been proved that the use of the XFEM can significantly improve convergence rates and accuracy for some problems. Furthermore, the XFEM suppresses the need to mesh and remesh the discontinuity surfaces, thus it can reduce the computational costs and projection errors available in conventional finite element method [68].

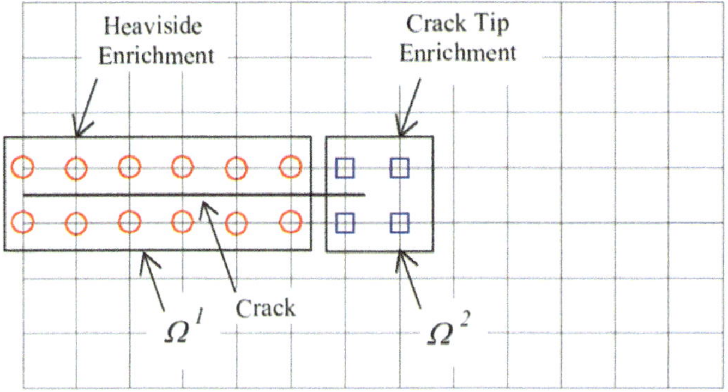

Fig. 2.27 Schematic of Heaviside and crack tip enrichments in the crack region [69]

As a numerical technique used by researchers to simulate crack extension in asphalt concretes, the XFEM has been shown to be very effective to characterize the discontinuous mechanical problems such as fractures. Indeed, the XFEM divides the numerical model into two sections. In the first section, the classical finite element meshes are generated for the un-cracked part of geometry; while, in the second section, the meshes defined in the cracked part of geometry are enriched by appropriate functions. Thus, the XFEM incorporates enrichment functions to solve fracture problems. The crack path is tracked with geometry-based level-set functions. As displayed in Fig. 2.27, the Heaviside and crack-tip enrichments are regarded in the crack region $\Omega = \Omega^1 + \Omega^2$. The enriched displacement function in XFEM is written as follows:

$$u^h(x) = \sum_{i \in I} u_i N_i + \sum_{j \in J} b_j N_j H(x) + \sum_{k \in K} N_k \left(\sum_{l=1}^{4} c_k^l F_l(x) \right) \tag{2.19}$$

N_i the traditional finite element shape functions at node i

u_i the traditional finite element DOFs (degrees of freedom) at node i

$H(x)$ the Heaviside (or jump) enrichment function

N_j the shape functions related to the Heaviside function at node j

b_j the additional DOFs related to the Heaviside enrichment functions (the circled nodes displayed in Fig. 2.27)

$F_l(x)$ the crack-tip enrichment function

N_k the shape function related to the enriched crack-tip function at node k

c_k the additional DOFs related to the elastic asymptotic crack-tip enrichment functions (the square nodes displayed in Fig. 2.27)

I the set of all nodes in the mesh

J the set of Heaviside enriched nodes whose shape functions are cut by the crack path

K the set of crack-tip enriched nodes whose shape functions are cut by the crack tip.

The Heaviside and crack-tip enrichments in a crack region are illustrated in Fig. 2.27. It is worth mentioning that the displacement functions are simplified to the traditional finite element method for the non-enriched regions, as follows:

$$u^h(x) = \sum_{i \in I} u_i N_i \tag{2.20}$$

For the crack region Ω^1 with Heaviside enriched elements, the displacement functions are given as:

$$u^h(x) = \sum_{i \in I} u_i N_i + \sum_{j \in J} b_j N_j H(x) \tag{2.21}$$

While, for the crack region Ω^2 with crack-tip enriched elements, the displacement functions are given as follows:

$$u^h(x) = \sum_{i \in I} u_i N_i + \sum_{k \in K} N_k \left(\sum_{l=1}^{4} c_k^l F_l(x) \right) \tag{2.22}$$

The Heaviside function $H(x)$ defined with a crack-surface level-set function $\phi(x)$ states the displacement jump between the crack surfaces within each element. It can be expressed as below:

$$H(x) = \text{sign}(\phi(x)) = \begin{cases} +1 : \phi(x) > 0 \ (\text{above the crack}) \\ -1 : \phi(x) < 0 \ (\text{below the crack}) \end{cases} \tag{2.23}$$

where, $\phi(x)$ is the crack-surface level-set function that states the signed normal distance of a Heaviside-enriched node with respect to the crack surface.

The crack-tip enrichment function represents the discontinuity at the crack tip within each element. It takes the form of four functions as follows:

$$[F_l(r, \theta), \ l = 1 - 4]$$
$$= \left[\sqrt{r} \sin \frac{\theta}{2}, \ \sqrt{r} \cos \frac{\theta}{2}, \ \sqrt{r} \sin \theta \sin \frac{\theta}{2}, \ \sqrt{r} \sin \theta \cos \frac{\theta}{2} \right] \tag{2.24}$$

where, r and θ are the polar coordinates in the local crack-tip coordinate system [69].

The XFEM was originally suggested by Belytschko and Black [70] and Dolbow [71], and later modified and applied to various crack problems by Daux et al. [72] and Sukumar et al. [73]. In recent years, the XFEM has been adopted to investigate crack growth in asphalt concretes. Several studies can be found in the literature on this issue [69, 74–78]. For example, Wang et al. [79] studied the crack growth in cracked SCB specimens made of rubber-modified asphalt concrete using the XFEM. It is noticed that, in comparison to the classical finite element method, the XFEM does not need to generate special meshes in the initial crack regions as the element edge and the set of crack-tip as the node. The XFEM regards the mesh and the crack independently, and does not generate dense meshes nearby the crack tip. It is not also required to construct new meshes along the crack growth path.

The base asphalt binder used by Wang et al. [79] had penetration grade of 60/80. Crumb rubber with the 40 mesh rubber fineness was used to produce the rubber modified asphalt mixtures. Furthermore, Dolerite and limestone were employed as the coarse aggregate (greater than 2.36 mm) and fine aggregate (smaller than 2.36 mm), respectively. The asphalt mixtures included continuous gradation with the 16 mm nominal maximum aggregate size and air void of 4%.

Selection of a proper criterion for the crack propagation is an important parameter in the numerical analysis using the XFEM. The J-integral criterion was used in this study. According to this criterion, the crack starts to propagate once the value of J reaches J_c (i.e. when J equals the critical value of J (i.e. J_c)). The value of J_c can be determined by the laboratory experiments. In order to obtain the value of J_c, the SCB specimens with three-different notch depths of 15, 20 and 25 mm were tested by loading with three-point bend fixture at $-15\,°C$. The SCB specimens were loaded with a constant cross-head deformation rate of 50 mm/min. The diameter d and thickness t of the SCB specimen were 150 and 25 mm, and the span value during the tests was assumed to be $0.8d$. The critical value of J-integral (i.e. J_c) was finally obtained as 4.97 kJ/m^2 for the rubber modified asphalt mixture. The fracture energy was also calculated as 4055.799, 3105.264 and 1571.303 N mm for the SCB specimens with notch depths of 15, 20 and 25 mm, respectively. Fracture energy and critical J-integral J_c not only are the necessary parameters for the crack propagation criterion, but also are a crucial step to perform the numerical simulation of the crack growth via the XFEM.

The SCB specimen with the notch depth of 25 mm and rubber content of 20% was chosen for laboratory experiment and numerical simulation via XFEM. Figure 2.28 plots variations of the load versus vertical displacement of the loading point (P-$LPVD$), in which the numerical results obtained from the XFEM exhibited good coincidence with the experimental results. The crack propagation path predicted by the XFEM was consistent with that of experiment. Thus, the results indicated the effectiveness of the numerical simulations by XFEM.

In another study performed by Lancaster et al. [80], an XFEM model was developed to predict the crack propagation using the SCB specimen. In this research, to capture the time dependent viscoelastic properties of asphalt within the XFEM model, a fractional viscoelastic element was employed instead of traditional springs and dashpots.

A binder with penetration grade of 60/70 and limestone aggregates were used for HMA preparation. In addition, they added a linear SBS polymer with three percentages (2.5, 5 and 7.5%) to the base binder to prepare the modified HMA mixtures. The SCB specimens with diameter of 150 mm and thickness of 50 mm were used for the fracture tests at 20 °C. The fracture tests were performed on the

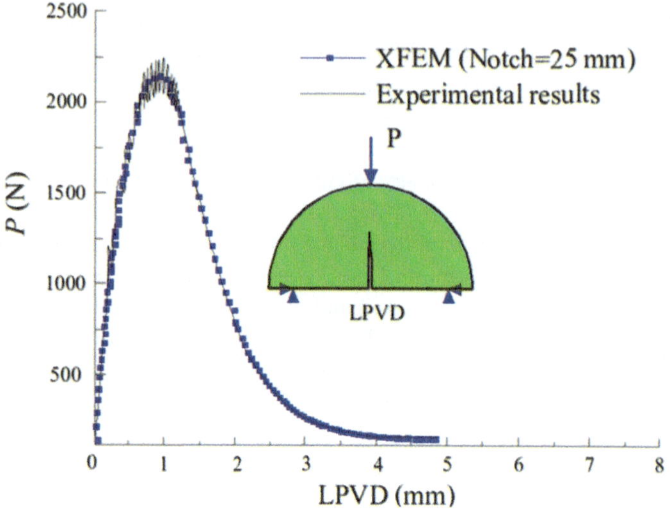

Fig. 2.28 Load versus LPVD, obtained from the numerical simulation and fracture test [79]

SCB specimens containing cracks with three different depths (i.e. 9, 19 and 29 mm) using a three-point bend fixture with a vertical displacement rate of 0.5 mm/min. In order to validate the use of XFEM, mode I stress intensity factors were obtained over a range of crack depth to radius ratios. For this purpose, a 3D model of the SCB test was generated in ABAQUS. The results (i.e. stress intensity factors) were then compared to the values obtained by Lim et al. [81]. A good agreement was observed between the results indicating that the XFEM can produce results with satisfactory accuracy.

A 2D model of the SCB was generated in ABAQUS to simulate the crack propagation under mode I loading. The models of XFEM with different crack depths of 9, 19 and 29 mm were constructed to match the laboratory test specimens. The creep data obtained from the small core testing were incorporated into the ABAQUS model to supply the viscoelastic material behavior. The fracture energy and peak cohesive stress for the material were then obtained by changing their values, by trial and error, until a match was achieved between the load–displacement curves obtained from the laboratory tests and XFEM simulations. The maximum principal stress was employed to consider the initiation of damage within the XFEM models. Damage is initiated when the maximum principal stress ratio (f) equals 1 [80].

$$f = \left(\frac{\langle \sigma_{max} \rangle}{\sigma_{max}^0} \right) \tag{2.25}$$

where, σ_{max}^0 is the maximum permitted principal stress, and σ_{max} is the maximum principal stress in the XFEM model. Meanwhile, $\langle \sigma_{max} \rangle = 0$ if $\sigma_{max} < 0$, and $\langle \sigma_{max} \rangle = \sigma_{max}$ if $\sigma_{max} \geq 0$. This is to ensure that the damage is not occurred by a compressive stress.

The results of the laboratory tests and XFEM models were then compared with each other. Based on these results, the XFEM model was reasonably able to predict the load-displacement curves obtained from the fracture experiments for all the HMA mixtures considered in [80].

2.7 Cohesive Zone Model

Asphalt mixtures are assumed to behave as brittle material at low temperatures (or at high loading rates), while they behave as quasi-brittle (or viscoelastic) material at intermediate and ambient temperatures (or at low loading rates). At low temperatures, linear elastic fracture mechanics (LEFM) approach can be reasonably employed to predict the onset of fracture (see e.g. [48, 49, 52, 59, 82–95]). For the cases that asphalt mixtures behave as quasi-brittle materials, the LEFM concept is no longer valid, and the fracture process zone ahead of the crack tip (as a non-elastic region) should be taken into account. The cohesive zone model (CZM), which considers the fracture process zone, has been found as an efficient approach to model crack initiation and propagation [96, 97].

The CZM was introduced by Dugdale [98] and Barenblatt [99]. Many researchers employed the CZM to investigate fracture behavior of different materials like metals, ceramics, polymers, geomaterials, etc. [100–103]. Hillerborg et al. [104] used the CZM concept for brittle materials. Needleman [105] pioneered the use of cohesive zone approach in finite element analysis for modeling the crack growth in ductile materials. Afterwards, many researchers employed the CZM to evaluate the fracture behavior of various materials (see e.g. [106–109]). A comprehensive review of the CZM can be found in [110].

The fundamental idea behind CZM is that as the distance between two atomic planes inside the material increases, the intensity of cohesion forces initially increases, and any further separation will give rise to an intensity drop. In the CZM, the traction between the atomic planes is attributed to the respective separation via a traction-separation law. The crack propagation can be simulated using the CZM as there is no traction between the crack surfaces. The crack can grow through the fracture process zone ahead of the crack which is controlled by a traction-separation law (see Fig. 2.29). The parameters t_n and δ_n shown in Fig. 2.29 refer to the traction and separation, respectively.

Choosing an appropriate traction-separation law is of an important step in the application of CZM [111]. Figure 2.30 shows commonly used traction-separation laws including polynomial, exponential, linear and bilinear relationships. More can be found in [112] on different traction-separation laws. Assuming a bilinear traction-separation law shown in Fig. 2.30d, three parameters including initial stiffness (E_0), cohesive strength (t_0) and cohesive energy (Γ_0) are required to be specified. Among the mentioned CZM parameters, the cohesive energy and the cohesive strength are physically meaningful. While, the initial stiffness should be selected great enough so that the CZM does not influence the overall compliance prior to damage initiation,

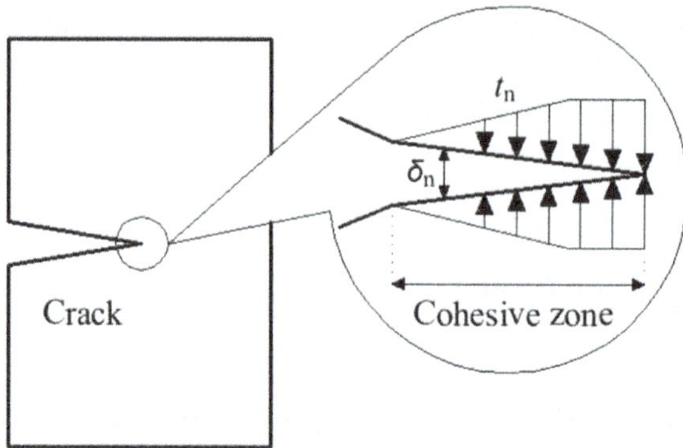

Fig. 2.29 Cohesive zone concept [46]

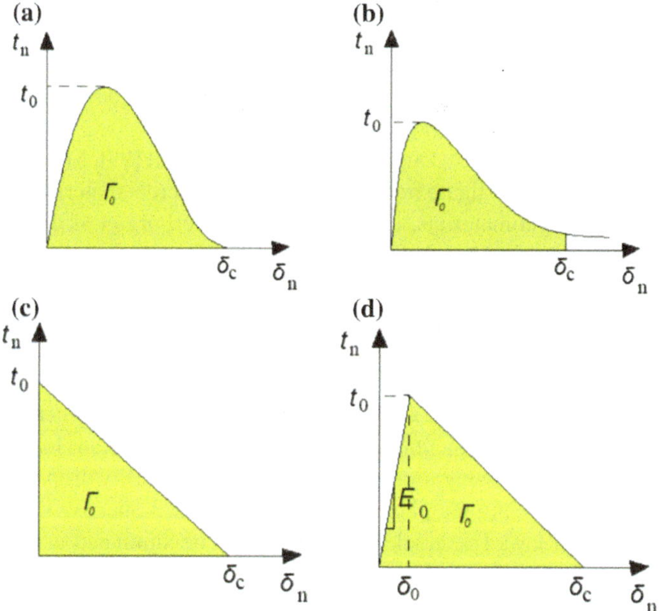

Fig. 2.30 Traction-separation laws, **a** polynomial, **b** exponential, **c** linear, **d** bilinear [46]

but from a numerical point of view, it's infinitely great value may result in numerical ill-conditioning.

The cohesive energy (Γ_0) is calculated as the area under the traction-separation curve. As seen in Fig. 2.30, the damage initiates as the traction t_n reaches its peak value (i.e. $t_n = t_0$) at the process zone, and by further increasing the separation, the traction decreases to zero at the critical separation value (i.e. $\delta_n = \delta_c$), which expresses propagation of the crack. One approach for characterizing the CZM parameters is to carry out numerical simulations and experiments, and then to compare the predicted and experimental results.

Several investigations can be also found in literature concerning application of the CZM on fracture of asphalt concretes (see e.g. [113–115]). Some of these studies are discussed herein in more detail.

2.7.1 Application of CZM in Finite Element Analysis of Asphalt Mixtures

Pirmohammad et al. [116] studied the effects of binder type and air void content on the CZM parameters. In order to determine the CZM parameters, experiments were performed on the SCB specimens made of various HMA mixtures under quasi-static loading. The specimen radius, the thickness and the pre-crack length were assumed as 75 mm, 32 mm and 20 mm, respectively. Five different types of HMA concretes with the NMAS of 19 mm were investigated in this study. Three of these HMA mixtures contained binders with penetration grades of 40, 60 and 85 (modified by 3.5% by weight of SBS), respectively and air void percentage of 4. Two of these mixtures contained 7% and 10% air void, respectively and binder with a penetration grade of 60. For the sake of convenience, HMA mixtures were designated as Bx, Vy in which B and V stood for the binder and air void, respectively, and the values of x and y represented the penetration grade of binder and the air void content.

To determine the tensile strength of HMA mixtures, the experiments were performed on un-cracked SCB specimens subjected to three-point bend loading at $-10\ °C$. The tensile strength tests were carried out in accordance with the recommendations of Molenaar et al. [65] such that the un-cracked SCB specimen was put on the fixed supports, and was then loaded by a crosshead displacement rate of 5 mm/min. The distance between the fixed supports (i.e. the span) was assumed to be 120 mm. The tensile strength of HMA mixtures was finally calculated from the following relation, as presented in Table 2.4.

$$\sigma_t = 4.263 \frac{P_{\max}}{2Rt} \tag{2.26}$$

where, $P_{\max}$ is the peak value of force recorded from experiments. R and t are the radius and thickness of the SCB specimen, respectively. Meanwhile, each experiment was repeated three times to obtain the average value of tensile strength.

In order to obtain the fracture energy of HMA mixtures investigated in this study at the temperature of $-10\ °C$, the cracked SCB specimens were employed. The SCB specimen was loaded symmetrically with a displacement rate of 0.2 mm/min using a fracture test set-up schematically shown in Fig. 2.31, and the load-$CMOD$ curve was recorded. The fracture energy G_f was then calculated as the area under the load-$CMOD$ curve divided by the fracture surface. The results are given in Table 2.5.

Table 2.4 The results of tensile strength tests for HMA mixtures [46]

Mixture	No.	P_{max} (kN)	σ_t (MPa)	Mixture	No.	P_{max} (kN)	σ_t (MPa)
B40, V4	1	6.10	5.42	B60, V4	1	7.09	6.30
	2	6.41	5.69		2	7.96	7.07
	3	5.93	5.27		3	7.48	6.64
	Ave.		5.46		Ave.		6.67
B85 and SBS, V4	1	9.25	8.22	B60, V7	1	6.17	5.48
	2	8.27	7.34		2	6.37	5.66
	3	8.74	7.76		3	5.32	4.72
	Ave.		7.77		Ave.		5.29
B60, V10	1	4.89	4.34				
	2	5.48	4.87				
	3	5.37	4.77				
	Ave.		4.66				

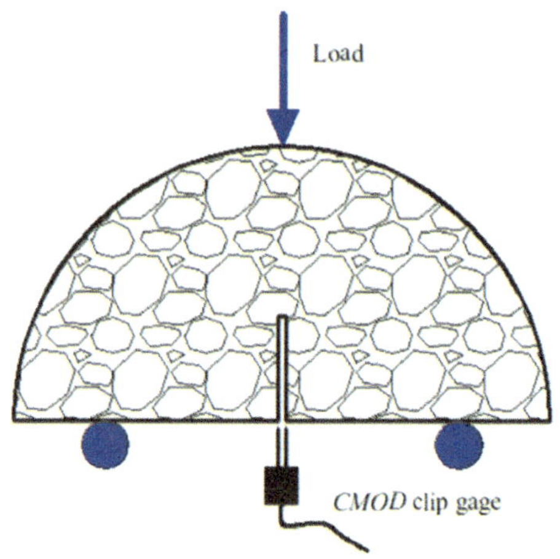

Fig. 2.31 Schematic of SCB test to obtain the fracture energy [46]

The procedure of characterizing the fracture energy of HMA mixtures would be comprehensively discussed in Chap. 5.

In the next step, the crack growth behavior of the HMA mixtures was simulated using the CZM concept. The 2D four-node cohesive elements (COH2D4) with bilinear traction–separation law were defined along the initial crack line (see Fig. 2.32). For other parts of the SCB specimen, four-node plane strain elements with the elastic modulus of $E = 14.2$ GPa and the Poisson's ratio $\upsilon = 0.35$ were defined. The bilinear traction-separation law can be expressed by parameters of the cohesive energy Γ_0 and the cohesive strength t_0. The cohesive energy was obtained as the area under the traction-separation curve, and the cohesive strength was calculated as the maximum value of traction (see Fig. 2.30d). In the CZM, damage initiates as the traction reaches

Table 2.5 The results of fracture energy G_f for HMA mixtures [46]

Mixture	No.	G_f (N/mm)	Mixture	No.	G_f (N/mm)
B40, V4	1	0.342	B60, V4	1	0.442
	2	0.317		2	0.382
	3	0.336		3	0.438
	Ave.	0.331		Ave.	0.421
B85 and SBS, V4	1	0.655	B60, V7	1	0.371
	2	0.555		2	0.357
	3	0.607		3	–
	Ave.	0.606		Ave.	0.364
B60, V10	1	0.236			
	2	0.323			
	3	0.336			
	Ave.	0.298			

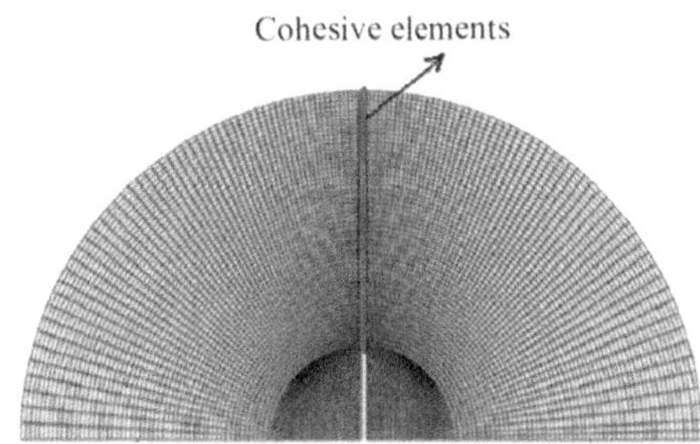

Fig. 2.32 Cohesive elements defined in the numerical analyses [46]

the cohesive strength, and then the material weakens according to the cohesive energy curve.

The cohesive zone parameters (i.e. Γ_0 and t_0) were obtained by calibrating the corresponding experimentally determined parameters of G_f and σ_t such that the model should successfully predict the experimental load-$CMOD$ response of the SCB specimen. Calibrated values of the cohesive energy and the cohesive strength are given in Table 2.6. Accordingly, both the cohesive energy and the cohesive strength increased as the binder in asphalt mixture softened. Additionally, the asphalt mixture modified with SBS exhibited the highest cohesive energy and cohesive strength values. This was because the application of binder with high penetration grade (i.e. PG 85) at low temperatures retarded the crack propagation due to local plastic deformations occurring at the crack tip. In other words, the asphalt mixture required more force and energy to fail. On the other hand, the addition of SBS to the base binder not only softened the resultant binder but also improved the adhesion between the aggregates and the binders [117, 118]. Consequently, both the softer binder (i.e. binder with penetration grade of 85) and SBS contributed to increase the values of cohesive energy

Table 2.6 The calibrated values of the cohesive energy and the cohesive strength for the HMA mixtures [46]

Mixture	Cohesive energy Γ_0 (N/mm)	Cohesive strength t_0 (MPa)
B40, V4	$1.33G_f = 1.33 \times 0.331 = 0.439$	$0.6\sigma_t = 0.6 \times 5.46 = 3.28$
B60, V4	$1.20G_f = 1.20 \times 0.421 = 0.510$	$0.8\sigma_t = 0.8 \times 6.67 = 5.34$
B85 and SBS, V4	$1.02G_f = 1.02 \times 0.606 = 0.618$	$0.8\sigma_t = 0.8 \times 7.77 = 6.22$
B60, V7	$1.30G_f = 1.30 \times 0.364 = 0.473$	$0.8\sigma_t = 0.8 \times 5.29 = 4.23$
B60, V10	$1.28G_f = 1.28 \times 0.298 = 0.382$	$0.6\sigma_t = 0.6 \times 4.66 = 3.12$

and cohesive strength. In addition, the higher air void content in asphalt mixture had deleterious effect on the cohesive parameters i.e. both the cohesive energy and the cohesive strength decreased as the air void content increased. This was because the air void weakened the structure of asphalt mixtures, and lower force or energy was therefore needed for fracturing the asphalt sample. In other words, air voids provided planes of weakness which reduced the resistance against the fracture propagation.

Figure 2.33 displays an example of the crack growth path in the SCB specimen obtained from the numerical simulation and experiment. The crack growth initiated from the initial crack tip, and extended straightly towards top point of the SCB specimen. Figure 2.34 shows the comparisons between the numerically predicted and experimentally obtained load-*CMOD* responses for various asphalt mixtures investigated [116].

The effects of different parameters including Γ_0, t_0, E and υ on the load-*CMOD* response of the SCB specimens were investigated in [49]. The material properties of the SCB specimen were regarded as those of B60, V4 (i.e. $\Gamma_0 = 0.510$ N/mm, $t_0 = 5.34$ MPa, $E = 14.2$ GPa and $\upsilon = 0.35$). Figure 2.35 shows the effects of the abovementioned parameters on the load-*CMOD* by considering the values of these parameters at three levels. As shown in Fig. 2.35a, the effect of cohesive energy was

(a)   **(b)**

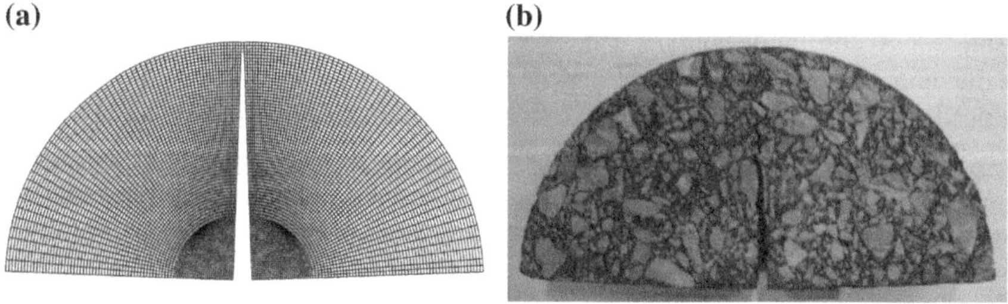

Fig. 2.33 Crack growth path in the SCB specimen obtained from **a** numerical simulation, **b** experiment [46]

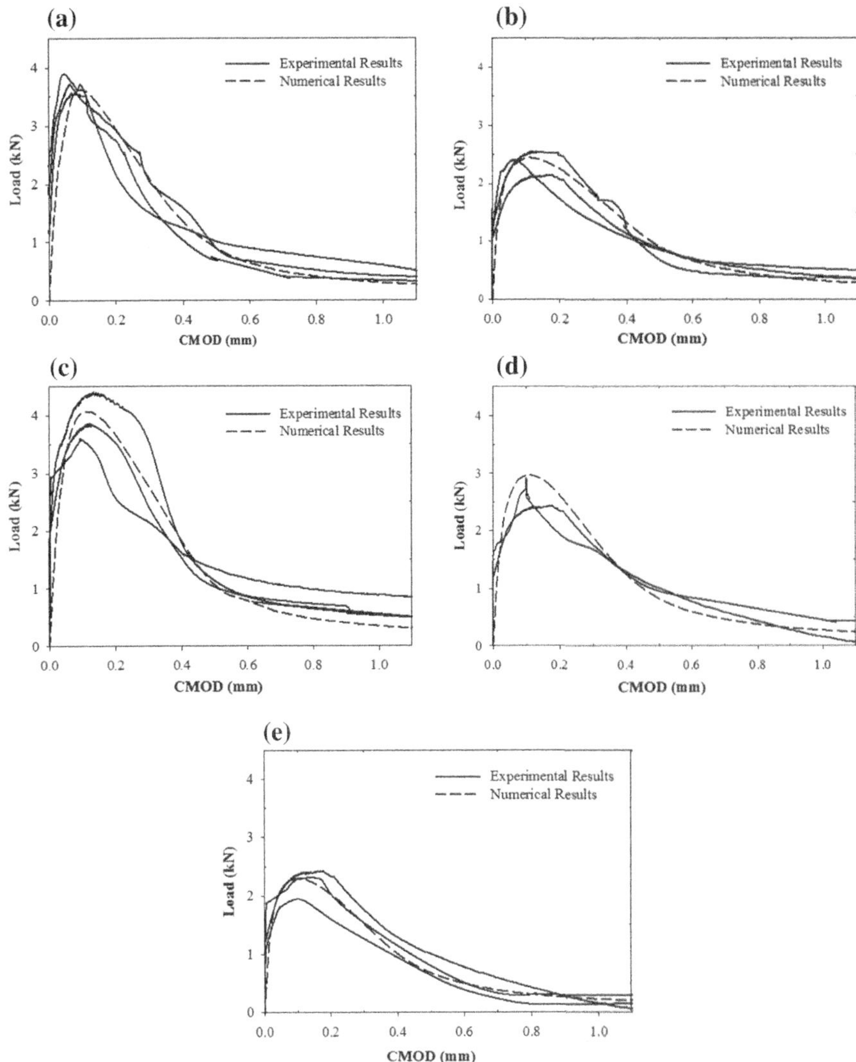

Fig. 2.34 Load-*CMOD* curves obtained from the experiments and numerical simulations for the asphalt mixtures of **a** B60, V4, **b** B40, V4, **c** B85 and SBS, V4, **d** B60, V7, **e** B60, V10 [46]

more pronounced at the softening part of the load-*CMOD* curve, and this curve was shifted towards the right side as the value of cohesive energy increased from $0.8\Gamma_0$ to $1.2\Gamma_0$. In addition, the peak value of load slightly increased. Figure 2.35b exhibits the effect of cohesive strength on the load-*CMOD* curve, in which the main effect of the cohesive strength was on the peak value of the load, and its value enhanced with increase in the cohesive strength.

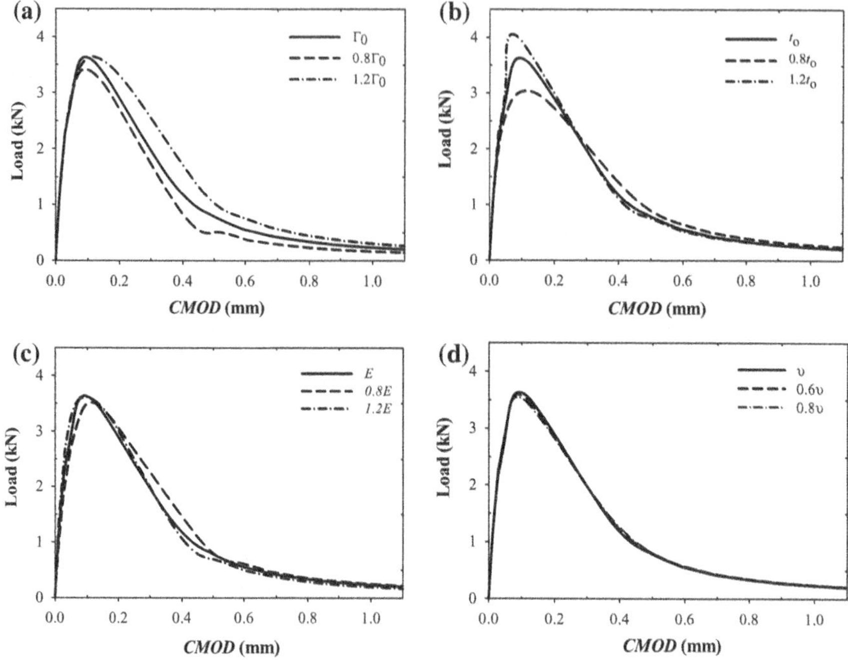

Fig. 2.35 The effect of different parameters including **a** cohesive energy, **b** cohesive strength, **c** elastic modulus, **d** Poisson's ratio on the load-*CMOD* curve [46]

The load-*CMOD* curve was slightly influenced by changing the value of elastic modulus, as shown in Fig. 2.35c. In addition, the load-*CMOD* curve was not affected by the Poisson's ratio (see Fig. 2.35d) considerably.

Li and Marasteanu [100] employed the CZM to simulate mode I fracture behavior of three asphalt mixtures at a low temperature using the SCB configuration with 150 mm in diameter and 25 mm in thickness. The SCB specimen containing an initial crack of 15 mm length in the middle was simply supported with a span of 120 mm and loaded from the top.

The mesh in the SCB specimen was built by 2D four-node quadrilateral elements. The cohesive elements were also defined over the initial un-cracked ligament in front of the initial crack (see Fig. 2.32).

Four material properties were required for incorporating into the numerical simulations as input data. The Young's modulus E, the tensile strength σ_t and the Poisson's ratio v were characterized from the Superpave indirect tension test (AASHTO T322-03, 2003), and the cohesive energy G_f was obtained from the SCB test. The input data used in the numerical simulations are summarized in Table 2.7. Meanwhile, some assumptions were made in the simulations for simplifications as follows:

- Change in the value of Poisson's ratio at low temperatures was negligible, so it was assumed to be 0.3 in the simulations.

Table 2.7 Material parameters used in numerical simulations [100]

Temperature	Material parameters	Asphalt mixture (PG)		
		58–28	58–34	58–40
−20 °C	Dynamic modulus E^* @ 1 Hz (GPa)	19.1	22.9	12.0
	Peak traction t_0 (MPa)	3.54	4.21	3.88
	Cohesive energy G_f (N/mm)	326	561	506
	Poisson's ratio	0.3	0.3	0.3

- The dynamic modulus E^* at 1 Hz was used as a surrogate Young's modulus in the simulations.
- The exponential traction-separation law was used in the simulations.
- The peak traction t_0 was assumed to be equal to the tensile strength σ_t of asphalt mixtures obtained from the Superpave indirect tension test (AASHTO T322-03, 2003), and would be calibrated later.
- The cohesive energy G_f was computed using the area under the load-*LLD* curve obtained from the SCB test.
- The critical separation was calculated by $\Gamma_0 = \exp(1)t_0\delta_c$.

The SCB specimen with material properties given in Table 2.7 was simulated, and the results of load-*LLD* were obtained. The numerical results were then compared with those of experiments to calibrate the cohesive parameters. A coefficient of 1 for the tensile strength (i.e. $t_0 = 1.0\sigma_t$) resulted in the best prediction for the 58–28 mixture, which contained a plain asphalt binder. For the 58–34 and 58–40 mixtures, which contained polymer-modified binders, a coefficient of 0.8 (i.e. $t_0 = 0.8\sigma_t$) resulted in the best prediction.

In another study, Song et al. [114] employed the cohesive elements to predict fracture behavior of disc-shaped compact tension (DC(T)) specimen. Figure 2.36 shows the DC(T) specimen used in this study with 150 mm in diameter, 145 mm

Fig. 2.36 Dimensions of the DC(T) specimen, and cohesive elements defined in the DC(T) [114]

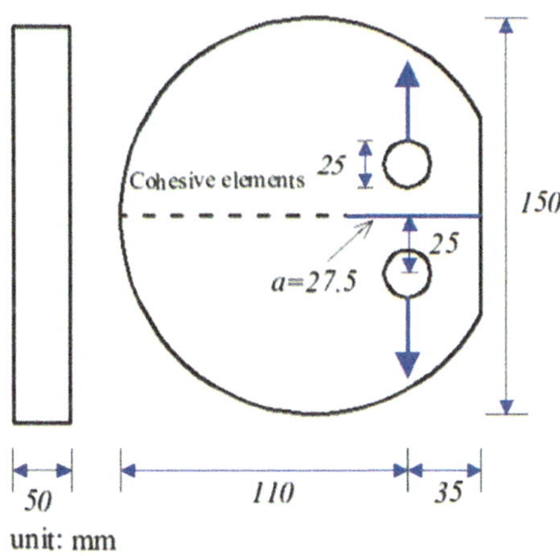

in length and 50 mm in thickness. The notch length was regarded as 27.5 mm. The specimen was loaded by pulling apart the pins inserted in the holes. The DC(T) specimen was meshed by using 2376 4-noded quadrilateral plane strain elements for the bulk elements and 884-noded linear elements for the cohesive elements. The cohesive elements of 1.0 mm size were regarded at the middle of the specimen and ahead of the initial notch. It is pointed out that the bilinear traction–separation law was used in the numerical simulations. The cohesive fracture energy of the mentioned DC(T) specimen at -10 °C and 1 mm/min *CMOD* loading rate was obtained $G_f = 324$ J/m^2, and the fracture strength at -10 °C was measured $\sigma_t = 3.58$ MPa. The modulus and Poisson's ratio were assumed as 11.9 GPa and 0.35, respectively. The cohesive parameters were calibrated by fitting the numerical results to the experimental ones in order to account for the differences between experiments and numerical simulations. By calibrating the cohesive parameters (i.e. $\Gamma_0 = 0.7 G_f = 0.7 \times 324$ J/m^2 and $t_0 = 0.95 \sigma_t = 0.95 \times 3.58$ MPa), the load-*CMOD* curve predicted by numerical simulation matched favorably with that of experimental results.

They used the calibrated cohesive parameters described above to simulate mixed mode I/II fracture in the SENB specimen. Unlike the DC(T) specimen where a crack path was predefined, in the simulation of mixed mode I/II fracture in the SENB specimen, the cohesive elements were defined over an area to allow cracks to propagate in any direction. Figure 2.37 displays dimensions of the SENB specimen, and cohesive elements were defined at the shaded region. A notch of 19 mm length was assumed at 65 mm relative to the center of the specimen. The shaded region was meshed by a regular pattern with 3-noded triangular cohesive elements, and 3-noded triangular elements were considered for bulk material. Meanwhile, the same cohesive fracture energy was regarded for mode I and mode II.

Figure 2.38a shows the deformed shape and crack path predicted by numerical simulation. Comparison of the crack paths obtained from experiments and numerical simulation is illustrated in Fig. 2.38b. Green and blue lines show the crack paths obtained from the experiments, and the red line state the crack path obtained from the numerical simulation, indicating favorable agreement.

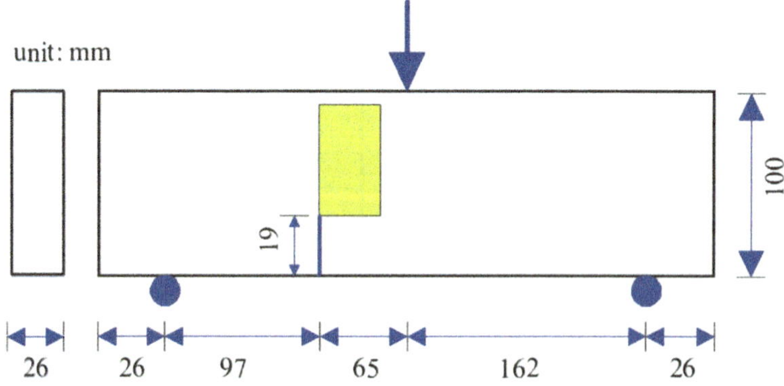

Fig. 2.37 Dimensions of the SENB specimen [114]

(a)

(b)

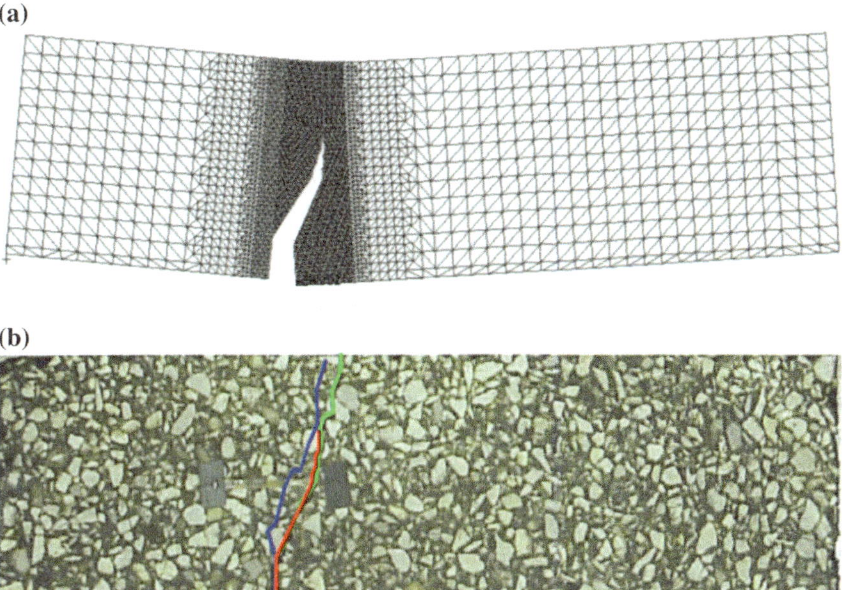

Fig. 2.38 Mixed mode I/II fracture of the SENB specimen, **a** deformed shape and crack path of numerical simulation, **b** comparison of the crack path obtained from the experiments (the green and blue lines) and numerical simulation (the red line) [114]

2.8 Summary

In this chapter, computational modeling of asphalt concrete with different fracture models was reviewed.

In the DEM, particles are bonded together at contact points and are separated by external forces. The DEM discretizes a material via rigid elements with simple shape. Each element interacts with neighboring elements based on appropriate interaction laws that are defined at contact points. Generally, two homogenous and heterogeneous discrete element models have been developed to simulate fracture behavior of asphalt concretes. In the homogenous models, the material behavior at the crack growth path is assumed cohesive, while it is considered elastic at other regions (i.e. the bulk material behaves in elastic). In other words, the effect of aggregate and mastic distributions is neglected in the homogenous models. In the heterogeneous discrete element models, an image processing technique is used for capturing the microstructure of asphalt mixtures. Digitalized specimen image with different aggregate size can be obtained by scanning laboratory asphalt specimens. Structure of asphalt concrete mixtures is made up of three different phases including aggregates, mastic (binder and fine sands), and the interface between aggregates and mastic. Material properties (i.e. elastic modulus, tensile strength and fracture energy) of these phases are required for numerical simulations, and must be determined separately. Aggregates are assumed to behave in linear elastic without softening. The mastic and the interface behave in nonlinear elastic with cohesive softening and adhesive softening, respectively. It is reminded that the tensile strength of asphalt concrete can be determined using the indirect tension test (IDT), and the fracture energy can be obtained by performing fracture tests on some specimens such as DC(T) and SENB. According to the literature, the DEM has been successfully used for modeling both the pure mode I and mixed mode I/II fracture of asphalt concretes.

In the subsequent section, the effect of different parameters including vehicle wheel position, horizontal load, elasticity and thickness of the road layers on the crack tip parameters (i.e. stress intensity factors, K_{I} and K_{II}, and T-stress) was studied using 2D finite element analyses. The road structure was assumed to have four layers of asphalt concrete, base, sub-base and sub-grade, and a top-down crack was regarded in the middle of the asphalt concrete layer. Based on the finite element results, the regions surrounding a top-down crack may be classified into three zones: (i) the first zone, which corresponds to the regions far enough from the crack, and only pure mode I deformation was observed at the crack tip, (ii) the second zone, which corresponds to the regions very close to the crack, and only pure mode II deformation was observed at the crack tip, and (iii) the third zone, which corresponds to the regions in between, and mixed I/II deformation mode took place at the crack tip. The results also showed that the horizontal load inducing from the vehicle acceleration or deceleration increased the crack tip parameters including K_{I}, K_{II} and T-stress significantly. Hence, the cracked asphalt concrete is subjected to critical loading condition where vehicles brake or accelerate (for example, at the intersections). As a result, an asphalt overlay with higher resistance against the crack growth must be used in these areas. According to the results of investigations performed on the effect of elasticity modulus of road layers, by increasing the asphalt concrete elasticity modulus, the crack tip parameters increased. Meanwhile, further reduction in the values of K_{I} was achieved by increasing the elasticity modulus of the beneath layers. As increase in the elasticity modulus occurred at lower layers, the elasticity modulus effect on the K_{II} and the T-stress diminished. For example, changing elasticity modulus of the sub-grade layer did not affect the K_{II} and the T-stress. It is also concluded that in order to avoid crack propagation at intersections, bus stations, etc., a material with lower elasticity modulus should be performed at the top layer of the road structure (for example, utilizing the modified asphalt concrete mixtures), and a material with higher elasticity modulus at beneath layers including base, sub-base and sub-grade. Finite element results for the effect of the thickness of road layers demonstrated that the crack tip parameters reduced as the thickness of road layers increased. Generally, by comparing the finite element results, it was observed that K_{I} and K_{II} were more affected by the thickness of the asphalt concrete than that of the base and sub-base layers.

Since 2D finite element modeling of road structure does not regard significant aspects of crack growth such as mode III deformations, so 3D finite element analysis can lead to more accurate and reliable results compared to the 2D one. According to the 3D finite element results, the opening deformation mode (i.e. mode I) was observed for a wide location of the wheels on the road; hence, this mode can be a major threat for the crack propagation and the deterioration of road structures. In addition to the opening mode, the in-plane and out-of-plane shear deformation modes (i.e. mode II and mode III) were also observed at the crack front. When the wheels located far from the crack plane, shear deformations were negligible compared to the total deformations at crack front. As the wheels approached the crack plane, the shear deformations became dominant relative to the opening deformation, and when the location of wheels was very close to the crack plane, the total deformation was nearly related to the shear modes. As a result, the shear deformation modes can play an important role in propagation of top-down crack under traffic loading.

Applicability of the extended finite element method (XFEM) was evaluated on the fracture behavior of asphalt concretes. The XFEM has been shown to be a very effective numerical simulation method to characterize the discontinuous mechanical problems such as crack extensions. Indeed, the XFEM divides the numerical model into two regions. In the first region, the classical finite element meshes are generated for the un-cracked part of geometry; while, in the second region, the meshes defined in the cracked part of geometry are enriched by appropriate functions. Thus, the XFEM incorporates enrichment functions to solve fracture problems. Some investigations were then reviewed, and the results demonstrated that the XFEM can be employed for predicting fracture behavior of asphalt concretes, in terms of both the force-displacement curve and crack growth path.

In the last section of this chapter, the application of the cohesive zone model (CZM) on fracture simulation of asphalt mixtures was discussed. To simulate the fracture behavior, four material properties including fracture energy, tensile strength (for the cohesive elements), Young's modulus and Poisson's ratio (for the bulk material) together with the load-$CMOD$ curve should be experimentally determined. In the next step, these parameters were defined in the simulations as input data, and the load-$CMOD$ curve was obtained. Finally, the fracture energy and tensile strength incorporated into the simulation were calibrated by fitting the numerically and experimentally obtained load-$CMOD$ curves to find the cohesive parameters including the cohesive energy and the cohesive strength.

References

1. Gdoutos EE (2018) Proceedings of the first international conference on theoretical, applied and experimental mechanics. Springer
2. Gdoutos EE (2019) Proceedings of the second international conference on theoretical, applied and experimental mechanics. Springer
3. González-Velázquez JL (2019) Mechanical behavior and fracture of engineering materials. Springer
4. González-Velázquez JL (2018) Fractography and failure analysis. Springer
5. Pirmohammad S, Hojjati Mengharpey M (2018) A new mixed mode I/II fracture test specimen: numerical and experimental studies. Theoret Appl Fract Mech 97:204–214

6. Wang L, Zhang B, Wang D, Yue Z (2007) Fundamental mechanics of asphalt compaction through FEM and DEM modeling. In: Analysis of asphalt pavement materials and systems: engineering methods, pp 45–63

7. Zelelew HM (2008) Simulation of the permanent deformation of asphalt concrete mixtures using discrete element method (DEM)

8. Chen J (2011) Discrete element method (DEM) analyses for hot-mix asphalt (HMA) mixture compaction

9. Kim H, Wagoner MP, Buttlar WG (2009) Numerical fracture analysis on the specimen size dependency of asphalt concrete using a cohesive softening model. Constr Build Mater 23(5):2112–2120

10. Yu H, Shen S (2013) A micromechanical based three-dimensional DEM approach to characterize the complex modulus of asphalt mixtures. Constr Build Mater 38:1089–1096

11. Cundall PA (1971) A computer model for simulating progressive, large-scale movement in blocky rock system. In: Proceedings of the international symposium on rock mechanics

12. Kim H, Wagoner MP, Buttlar WG (2009) Micromechanical fracture modeling of asphalt concrete using a single-edge notched beam test. Mater Struct 42(5):677

13. Cundall PA, Hart RD (1995) Numerical modeling of discontinua. In: Analysis and design methods. Elsevier, pp 231–243

14. Kawai T (1978) New discrete models and their application to seismic response analysis of structures. Nucl Eng Des 48(1):207–229

15. Hocking G (1992) The discrete element method for analysis of fragmentation of discontinua. Eng Comput 9(2):145–155

16. Bolander J Jr, Saito S (1997) Discrete modeling of short-fiber reinforcement in cementitious composites. Adv Cem Based Mater 6(3–4):76–86

17. d'Addetta GA (2004) Discrete models for cohesive frictional materials

18. Tan Y, Yang D, Sheng Y (2009) Discrete element method (DEM) modeling of fracture and damage in the machining process of polycrystalline SiC. J Eur Ceram Soc 29(6):1029–1037

19. Kim H, Wagoner MP, Buttlar WG (2008) Simulation of fracture behavior in asphalt concrete using a heterogeneous cohesive zone discrete element model. J Mater Civ Eng 20(8):552–563

20. Sadd MH, Dai Q, Parameswaran V, Shukla A (2004) Microstructural simulation of asphalt materials: modeling and experimental studies. J Mater Civ Eng 16(2):107–115

21. Kim H, Buttlar WG (2009) Discrete fracture modeling of asphalt concrete. Int J Solids Struct 46(13):2593–2604

22. Schlangen E, Garboczi E (1997) Fracture simulations of concrete using lattice models: computational aspects. Eng Fract Mech 57(2–3):319–332

23. Gao H, Yang X, Zhang C (2015) Experimental and numerical analysis of three-point bending fracture of pre-notched asphalt mixture beam. Constr Build Mater 90:1–10

24. Kleemans CP, Zuidema J, Krans RL, Molenaar JMM, Tolman F (1997) Fatigue and creep crack growth in fine-sand asphalt materials. J Test Eval 25(4):424–428

25. Molenaar AAA (2001) Prediction of fatigue cracking in asphalt pavements. Transp Res Rec 2007:155–162

26. Garzon J, Duarte CA, Buttlar WG (2010) Computational simulations of a full-scale reflective cracking test. In: 2010 FAA worldwide airport technology transfer conference federal aviation administration, Atlantic City, New Jersey, USA. American Association of Airport Executives

27. Molenaar AAA (1984) Fatigue and reflection cracking due to traffic loads. J Assoc Asphalt Paving Technol 53:440–474

28. Seung WL, Jae MB, Seung HH, Shelley MS (2007) Evaluation of optimum rubblized depth to prevent reflection cracks. J Transp Eng 133(6):355–361

29. Al-Qadi I, Wang H (2009) Evaluation of pavement damage due to new tire designs. Research report ICT-09-048

30. Emery J (2007) Mitigation of asphalt pavement top-down cracking. In: 4th international SIIV congress, Palermo, Italy

31. Scarpas A, De Bondt A, Molenaar A, Gaarkeuken G (1996) Finite elements modelling of cracking in pavements. In: RILEM proceedings. Chapman & Hall, pp 82–91

32. Zhou F, Hu S, Hu X, Scullion T (2009) Mechanistic-empirical asphalt overlay thickness design and analysis system. Report No. FHWA/TX-09/0-5123-3. Texas Transportation Institute, The Texas A&M University System, College Station, Texas
33. Loizos A, Partl MN, Scarpas T, Al-Qadi IL (2009) Modeling of top-down cracking (TDC) propagation in asphalt concrete pavements using fracture mechanics theory. In: Advanced testing and characterization of bituminous materials, two volume set. CRC Press, pp 713–724
34. Luo H, Zhu H-P, Miao Y, Chen C-Y (2010) Simulation of top-down crack propagation in asphalt pavements. J Zhejiang Univ Sci A 11(3):223–230
35. Hyunwook K, Buttlar WG, Chou KF (2010) Mesh-independent fracture modeling for overlay pavement system under heavy aircraft gear loadings. J Transp Eng 136(4):370–378
36. Yang B, Huang QS, Xiong B, Xu H (2011) Analysis on the effect of parameters variation on dynamic stress intensity factor of asphalt overlay reflective crack. In: Applied mechanics and materials. Trans Tech Publ, pp 1092–1096
37. Ceylan H, Gopalakrishnan K, Lytton RL (2010) Neural networks modeling of stress growth in asphalt overlays due to load and thermal effects during reflection cracking. J Mater Civ Eng 23(3):221–229
38. Shalaby A, Abd El Halim A, Easa S (1996) Influence of thermal stresses on construction-induced crack. In: Third conference on reflection cracking, pp 30–39
39. Suo Z (2012) Evaluation of asphalt concrete pavement service life using 3D nonlinear finite element analysis and nonlinear fatigue damage model. The Hong Kong Polytechnic University
40. Su K, Sun L, Hachiya Y, Maekawa R (2008) Analysis of shear stress in asphalt pavements under actual measured tire-pavement contact pressure. In: Proceedings of the 6th ICPT, Sapporo, Japan, pp 11–18
41. Ameri M, Mansourian A, Khavas MH, Aliha M, Ayatollahi M (2011) Cracked asphalt pavement under traffic loading—a 3D finite element analysis. Eng Fract Mech 78(8):1817–1826
42. Pirmohammad S, Majd-Shokorlou Y (2020) Finite element analysis of road structure containing a top-down crack within asphalt concrete layer. J Cent South Univ 27:242–255
43. Liang RY, Zhu JX (1995) Dynamic analysis of infinite beam on modified Vlasov subgrade. J Transp Eng 121(5):434–442
44. Novak M, Birgisson B, Roque R (2003) Near-surface stress states in flexible pavements using measured radial tire contact stresses and ADINA. Comput Struct 81:859–870
45. Uddin W, Zhang D, Fernandez F (1994) Finite element simulation of pavement discontinuities and dynamic load response. Transp Res Rec 1448:100–106
46. Pirmohammad S (2013) Study of mixed mode I/II fracture behavior of asphaltic materials at low temperatures. Ph.D. dissertation, Mechanical Engineering, Iran University of Science and Technology
47. Timm D, Birgisson B, Newcomb D (1998) Development of mechanistic-empirical pavement design in Minnesota. Transp Res Rec 1629(1):181–188
48. Ameri M, Mansourian A, Pirmohammad S, Aliha M, Ayatollahi M (2012) Mixed mode fracture resistance of asphalt concrete mixtures. Eng Fract Mech 93:153–167
49. Ayatollahi MR, Pirmohammad S (2013) Temperature effects on brittle fracture in cracked asphalt concretes. Struct Eng Mech 45(1):19–32
50. Pirmohammad S, Ayatollahi M (2015) Asphalt concrete resistance against fracture at low temperatures under different modes of loading. Cold Reg Sci Technol 110:149–159
51. Pirmohammad S, Kiani A (2016) Impact of temperature cycling on fracture resistance of asphalt concretes. Comput Concr 17(4):541–551
52. Pirmohammad S, Kiani A (2016) Study on fracture behavior of HMA mixtures under mixed mode I/III loading. Eng Fract Mech 153:80–90
53. Aliha M, Sarbijan M (2016) Effects of loading, geometry and material properties on fracture parameters of a pavement containing top-down and bottom-up cracks. Eng Fract Mech 166:182–197
54. Rostami M (2015) Fracture behavior of road structure containing a bottom-up crack. M.Sc. dissertation, University of Mohaghegh Ardabili

55. Ayatollahi MR, Pirmohammad S, Sedighiani K (2014) Three-dimensional finite element modeling of a transverse top-down crack in asphalt concrete. Comput Concr 13(4):569–585
56. Jung D, Vinson TS (1993) Low temperature cracking resistance of asphalt concrete mixtures (with discussion). J Assoc Asphalt Paving Technol 62
57. Zubeck H, Zeng H, Vinson T, Janoo V (1996) Field validation of thermal stress restrained specimen test: six case histories. Transp Res Rec 1545:67–74
58. Li XJ, Marasteanu M (2010) Using semi circular bending test to evaluate low temperature fracture resistance for asphalt concrete. Exp Mech 50(7):867–876
59. Pirmohammad S, Bayat A (2017) Fracture resistance of HMA mixtures under mixed mode I/III loading at different subzero temperatures. Int J Solids Struct 120:268–277
60. Kim KW, Kweon SJ, Doh YS, Park T-S (2003) Fracture toughness of polymer-modified asphalt concrete at low temperatures. Can J Civ Eng 30(2):406–413
61. Casey D, McNally C, Gibney A, Gilchrist MD (2008) Development of a recycled polymer modified binder for use in stone mastic asphalt. Resour Conserv Recycl 52(10):1167–1174
62. Buttlar WG, Armando DC, Garzon J (2004) Development of new methodologies for mechanistic design of asphalt overlays
63. Kim KW, El Hussein M (1997) Variation of fracture toughness of asphalt concrete under low temperatures. Constr Build Mater 11(7–8):403–411
64. Molenaar J, Molenaar A (2000) Fracture toughness of asphalt in the semi-circular bend test. In: 2nd Eurasphalt and Eurobitume congress, Barcelona, Spain, 20–22 Sept 2000
65. Molenaar J, Liu X, Molenaar A (2003) Resistance to crack-growth and fracture of asphalt mixture. In: 6th international RILEM symposium, Zurich, Switzerland, 14–16 Apr 2003
66. Wagnoner M, Buttlar WG, Paulino G (2005) Disk-shaped compact tension test for asphalt concrete fracture. Exp Mech 45(3):270–277
67. Pirmohammad S, Ayatollahi M (2014) Fracture resistance of asphalt concrete under different loading modes and temperature conditions. Constr Build Mater 53:235–242
68. Datta D (2013) Introduction to extended finite element (XFEM) method. arXiv preprint arXiv:1308.5208
69. Ng K, Dai Q (2011) Investigation of fracture behavior of heterogeneous infrastructure materials with extended-finite-element method and image analysis. J Mater Civ Eng 23(12):1662–1671
70. Belytschko T, Black T (1999) Elastic crack growth in finite elements with minimal remeshing. Int J Numer Meth Eng 45(5):601–620
71. Dolbow JE (2000) An extended finite element method with discontinuous enrichment for applied mechanics
72. Daux C, Moës N, Dolbow J, Sukumar N, Belytschko T (2000) Arbitrary branched and intersecting cracks with the extended finite element method. Int J Numer Meth Eng 48(12):1741–1760
73. Sukumar N, Moës N, Moran B, Belytschko T (2000) Extended finite element method for three-dimensional crack modelling. Int J Numer Meth Eng 48(11):1549–1570
74. Mahmoud E, Saadeh S, Hakimelahi H, Harvey J (2014) Extended finite-element modelling of asphalt mixtures fracture properties using the semi-circular bending test. Road Mater Pavement Des 15(1):153–166
75. Wang X, Li K, Zhong Y, Xu Q, Li C (2018) XFEM simulation of reflective crack in asphalt pavement structure under cyclic temperature. Constr Build Mater 189:1035–1044
76. Qian ZD, Hu J (2012) Fracture properties of epoxy asphalt mixture based on extended finite element method. J Cent South Univ 19(11):3335–3341
77. Im S, Ban H, Kim Y-R (2014) Characterization of mode-I and mode-II fracture properties of fine aggregate matrix using a semicircular specimen geometry. Constr Build Mater 52:413–421
78. Ban H, Im S, Kim Y-R (2015) Mixed-mode fracture characterization of fine aggregate mixtures using semicircular bend fracture test and extended finite element modeling. Constr Build Mater 101:721–729

79. Wang H, Zhang C, Yang L, You Z (2013) Study on the rubber-modified asphalt mixtures' cracking propagation using the extended finite element method. Constr Build Mater 47:223–230
80. Lancaster I, Khalid H, Kougioumtzoglou I (2013) Extended FEM modelling of crack propagation using the semi-circular bending test. Constr Build Mater 48:270–277
81. Lim I, Johnston I, Choi S (1993) Stress intensity factors for semi-circular specimens under three-point bending. Eng Fract Mech 44(3):363–382
82. Tekalur SA, Shukla A, Sadd M, Lee KW (2009) Mechanical characterization of a bituminous mix under quasi-static and high-strain rate loading. Constr Build Mater 23(5):1795–1802
83. Li X, Marasteanu M (2004) Evaluation of the low temperature fracture resistance of asphalt mixtures using the semi circular bend test (with discussion). J Assoc Asphalt Paving Technol 73
84. Aliha M, Fazaeli H, Aghajani S, Nejad FM (2015) Effect of temperature and air void on mixed mode fracture toughness of modified asphalt mixtures. Constr Build Mater 95:545–555
85. Behbahani H, Aliha M, Reza M, Fazaeli H, Aghajani S (2013) Experimental fracture toughness study for some modified asphalt mixtures. In: Advanced materials research. Trans Tech Publ, pp 337–344
86. Pirmohammad S, Bayat A (2016) Characterizing mixed mode I/III fracture toughness of asphalt concrete using asymmetric disc bend (ADB) specimen. Constr Build Mater 120:571–580
87. Pirmohammad S, Kiani A (2016) Effect of temperature variations on fracture resistance of HMA mixtures under different loading modes. Mater Struct 49(9):3773–3784
88. Pirmohammad S, Yousefi A, Sobhi S, Vaseghi Z (2020) Fracture strength of warm mix asphalt (WMA) concretes containing reclaimed asphalt pavement (RAP). J Cent South Univ (under review)
89. Pirmohammad S, Majd-Shokorlou Y, Amani B (2020) Experimental investigation of fracture properties of asphalt mixtures modified with nano Fe_2O_3 and carbon nanotubes. Road Mater Pavement Des 1–23 (in press)
90. Pirmohammad S, Majd-Shokorlou Y, Amani B (2019) Fracture resistance of HMA mixtures modified with nanoclay and nano-Al_2O_3. J Test Eval 47(5):3289–3308
91. Pirmohammad S, Khanpour M (2020) Fracture strength of warm mix asphalt (WMA) concretes modified with crumb rubber subjected to variable temperatures. Road Mater Pavement Des (in press)
92. Pirmohammad S, Majd-Shokorlou Y, Amani B (2020) Influence of natural fibers (kenaf and goat wool) on mixed mode I/II fracture strength of asphalt mixtures. Constr Build Mater 239 (in press)
93. Pirmohammad S, Majd-Shokorlou Y, Amani B (2020) Laboratory investigations on fracture toughness of asphalt concretes reinforced with carbon and kenaf fibers. Eng Fract Mech 226 (in press)
94. Pirmohammad S, Shabani H (2019) Mixed mode I/II fracture strength of modified HMA concretes subjected to different temperature conditions. J Test Eval 47(5):3355–3371
95. Pirmohammad S, Amani B, Majd-Shokorlou Y (2020) The effect of basalt fibers on fracture toughness of asphalt mixture. Fatigue Fract Eng Mater Struct (in press)
96. Xu XP, Needleman A (1994) Numerical simulations of fast crack growth in brittle solids. J Mech Phys Solids 42(9):1397–1434
97. Camacho GT, Ortiz M (1996) Computational modelling of impact damage in brittle materials. Int J Solids Struct 33(20–22):2899–2938
98. Dugdale DS (1960) Yielding of steel sheets containing slits. J Mech Phys Solids 8(2):100–104
99. Barenblatt GI (1962) The mathematical theory of equilibrium cracks in brittle fracture. In: Advances in applied mechanics. Elsevier, pp 55–129
100. Li X, Marasteanu MO (2005) Cohesive modeling of fracture in asphalt mixtures at low temperatures. Int J Fract 136(1–4):285–308
101. Koloor S, Abdullah M, Tamin M, Ayatollahi M (2019) Fatigue damage of cohesive interfaces in fiber-reinforced polymer composite laminates. Compos Sci Technol 183:107779

102. Torabi A, Majidi H, Ayatollahi M (2019) Fracture study in notched graphite specimens subjected to mixed mode I/II loading: application of XFEM based on the cohesive zone model. Theoret Appl Fract Mech 99:60–70
103. Khoramishad H, Akbardoost J, Ayatollahi M (2014) Size effects on parameters of cohesive zone model in mode I fracture of limestone. Int J Damage Mech 23(4):588–605
104. Hillerborg A, Modéer M, Petersson P-E (1976) Analysis of crack formation and crack growth in concrete by means of fracture mechanics and finite elements. Cem Concr Res 6(6):773–781
105. Needleman A (1990) An analysis of decohesion along an imperfect interface. Int J Fract 42:21–40
106. Elices M, Rocco C, Roselló C (2009) Cohesive crack modelling of a simple concrete: experimental and numerical results. Eng Fract Mech 76(10):1398–1410
107. Khoramishad H, Crocombe A, Katnam K, Ashcroft I (2011) Fatigue damage modelling of adhesively bonded joints under variable amplitude loading using a cohesive zone model. Eng Fract Mech 78(18):3212–3225
108. Carrier B, Granet S (2012) Numerical modeling of hydraulic fracture problem in permeable medium using cohesive zone model. Eng Fract Mech 79:312–328
109. Maiti S, Geubelle PH (2005) A cohesive model for fatigue failure of polymers. Eng Fract Mech 72(5):691–708
110. Bazant ZP, Planas J (1997) Fracture and size effect in concrete and other quasibrittle materials. CRC Press
111. Cornec A, Scheider I, Schwalbe K-H (2003) On the practical application of the cohesive model. Eng Fract Mech 70(14):1963–1987
112. Chandra N, Li H, Shet C, Ghonem H (2002) Some issues in the application of cohesive zone models for metal–ceramic interfaces. Int J Solids Struct 39(10):2827–2855
113. de Souza F, Soares J, Allen D, Evangelista F Jr (1891) Model for predicting damage evolution in heterogeneous viscoelastic asphaltic mixtures. Transp Res Rec 2004:131–139
114. Song SH, Paulino GH, Buttlar WG (2006) A bilinear cohesive zone model tailored for fracture of asphalt concrete considering viscoelastic bulk material. Eng Fract Mech 73(18):2829–2848
115. Li X, Marasteanu M (2010) The fracture process zone in asphalt mixture at low temperature. Eng Fract Mech 77(7):1175–1190
116. Pirmohammad S, Khoramishad H, Ayatollahi M (2015) Effects of asphalt concrete characteristics on cohesive zone model parameters of hot mix asphalt mixtures. Can J Civ Eng 43(3):226–232
117. Bhurke A, Shin E, Drzal L (1997) Fracture morphology and fracture toughness measurement of polymer-modified asphalt concrete. Transp Res Rec 1590:23–33
118. Champion L, Gerard J, Planche J, Martin D, Anderson D (1999) Evaluation of the low-temperature fracture properties of modified binders. Relationship with their micromorphology. In: Workshop briefing, performance related properties for bituminous binders, Eurobitume Workshop, Luxembourg

Chapter 3
Fracture Behavior of HMA Concretes at Low Temperatures

Abstract This chapter discusses on the fracture behavior of hot mix asphalt (HMA) concretes. Procedure of performing fracture tests by using different test specimens including SENB, DC(T), SCB, etc. are described. These specimens are able to produce various combinations of mode I and mode II deformation at the crack tip. The effects of different parameters including aggregate type (e.g., granite, limestone, etc.), aggregate gradation, air void content, binder (i.e., binder type and binder content), temperature, nanomaterials (e.g., nanoclay, carbon nanotubes, etc.), fibers (e.g., kenaf, basalt, carbon, etc.) and additives (e.g., crumb rubber, SBS, etc.) on the fracture toughness of HMA mixtures are then discussed by reviewing investigations performed on these issues. Fracture behavior of HMA concretes subjected to mixed mode I/III and pure mode III loadings is finally evaluated.

3.1 Introduction

Hot mix asphalt (HMA) concretes are considered as a composite material making up of aggregates, binders, and air voids. Binder constitutes about 5% by weight and 15% by volume of a HMA concrete and plays a significant role in the performance of HMA concretes. Binder is known to be temperature-dependent material. By decreasing the temperature particularly in cold climate regions, the binder in HMA concrete becomes brittle and loses its ductility. It improves the resistance of asphalt concrete against rutting, but cracking becomes the more likely mode of deterioration. Some researchers reported that binders of different penetration grades show different fracture strengths, and therefore, an appropriate binder type must be used for preparation of HMA concretes serving at different temperature conditions. Meanwhile, aggregates (as a major component used in the preparation of HMA concretes) and air voids can also affect cracking behavior of the asphalt concretes. In addition, previous investigations have exhibited that temperature variations can also influence the fracture behavior of HMA concretes to some extent.

There are extensive studies on mode I fracture behavior of HMA concretes, and many efforts have been also made in recent years to investigate other crack growth mechanisms including mixed mode I/II and I/III. It is pointed out that researchers

© Springer Nature Switzerland AG 2020

S. Pirmohammad and M. R. Ayatollahi, *Fracture Behavior of Asphalt Materials*, Structural Integrity 14, https://doi.org/10.1007/978-3-030-39974-0_3

employed several test specimens (for example, single edge notched beam (SENB), disk-shaped compact tension (DC-T), semicircular bend (SCB), disk bend, etc.) to carry out the fracture experiments under different loading modes including pure mode I, pure mode II, pure mode III, mixed mode I/II, and mixed mode I/III.

It is well known that the addition of polymers to binders can improve the performance of HMA concretes to protect the top layer of road structures against cracking and to provide good ride quality. Polymers play an important role in the pavement industry. They are among the most technically advanced binder modifiers currently available. The degree of modification by polymers depends on the polymer property, polymer content, and nature of the binder. In order to improve performance of HMA mixtures, many additives such as styrene-butadiene-styrene (SBS), crumb rubber (CR), ethylene vinyl acetate (EVA), and polyethylene (PE) have been used in the past. In addition, several investigations have been carried out on the use of nanomaterials and fibers to study their effect on fracture behavior of asphalt mixtures.

The abovementioned issues affecting the fracture behavior of HMA concretes are comprehensively discussed in this chapter.

3.2 Mixed Mode I/II Fracture Toughness

In order to measure fracture toughness of asphalt mixtures, the following three steps are generally performed.

(i)   Selection of suitable test specimen,
(ii)   Numerical analysis on the test specimen to calculate the geometry factors,
(iii) Conducting the experiments to calculate the fracture toughness.

These steps are discussed in the subsequent sections in detail.

3.2.1 Test Specimen

Different test specimens have been used by researchers to investigate fracture behavior of asphalt mixtures. Figure 3.1 shows the SENB [1–7], DC(T) [8, 9], and SCB [5, 10–15] specimens, which are often used for performing the fracture tests under mode I loading. As shown in Fig. 3.1, the load on both the SENB and SCB specimens is applied by three-point bend fixture. For the DC(T) specimen, arms are put into two holes and are then pulled apart.

From the specimens illustrated in Fig. 3.1, the SENB and SCB can be used for performing the fracture tests under mixed mode I/II loading. For example, Artamendi and Khalid [5], Kim et al. [16], Xie et al. [17], Gao et al. [18], and Song et al. [19] have used the SENB specimen to study mixed mode I/II fracture behavior of asphalt mixtures by translating the crack away from the middle of the specimen (see Fig. 3.2a). Braham and Buttlar [20] used another configuration of beam test

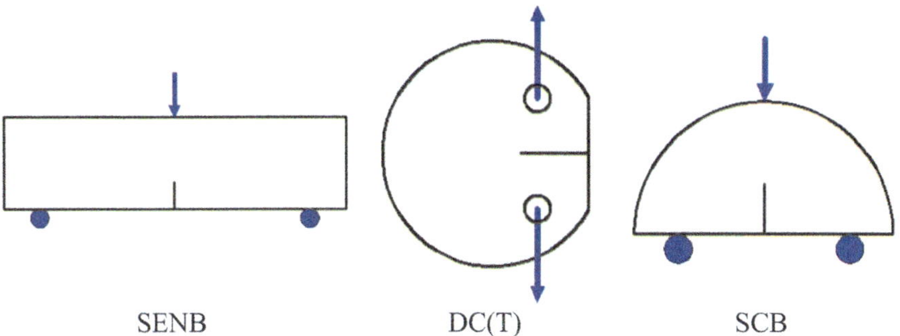

Fig. 3.1 Popular specimens used for mode I fracture test on asphalt mixtures

illustrated in Fig. 3.2b to explore pure mode II fracture of asphalt mixtures. This specimen included double notches created at the top and bottom of the beam and a four-point fixture was used for loading the specimen; meanwhile, the specimen was fixed at the two ends for avoiding its bending. Some researchers such as Artamendi and Khalid [5] and Im et al. [21] have employed the SCB specimen to investigate mixed mode I/II fracture behavior of asphalt mixtures by rotating the initial crack (see Fig. 3.3). It is pointed out that Ayatollahi and his coworkers [14, 22–28] have also used the SCB specimen shown in Fig. 3.3 to explore mixed mode I/II fracture behavior of different materials like rocks, polymers, and so on. According to Ameri et al. [29], carving an inclined crack in the SCB specimens made of asphalt mixtures was often difficult and sometimes led to breakage of aggregates at the center of the SCB specimen. Hence, the SCB specimens shown in Fig. 3.4 have been used by many researchers, in which the initial crack is normal to the flat side of the SCB specimen. For the SCB specimen shown in Fig. 3.4a, the mode mixity is controlled by changing the value of L (*i.e.*, by translating the crack), while positions of the bottom supports are fixed at S_1 and S_2. For the SCB specimen shown in Fig. 3.4b, the mode mixity is controlled by changing the positions of the bottom supports (*i.e.*, S_1 and S_2), while the crack is located in the middle of the specimen. According to Ameri et al. [29], both of the SCB specimens shown in Fig. 3.4 had the potential for investigating the fracture behavior of asphalt mixtures; however, the SCB (type I) shown in Fig. 3.4a was found to be more suitable.

3.2.2 *Numerical Analysis of SCB Specimen*

Mixity parameter M^e expresses the relative contributions of mode I and mode II and can be written as follows [30]:

$$M^e = \frac{2}{\pi} \tan^{-1}\left(\frac{K_I}{K_{II}}\right) \tag{3.1}$$

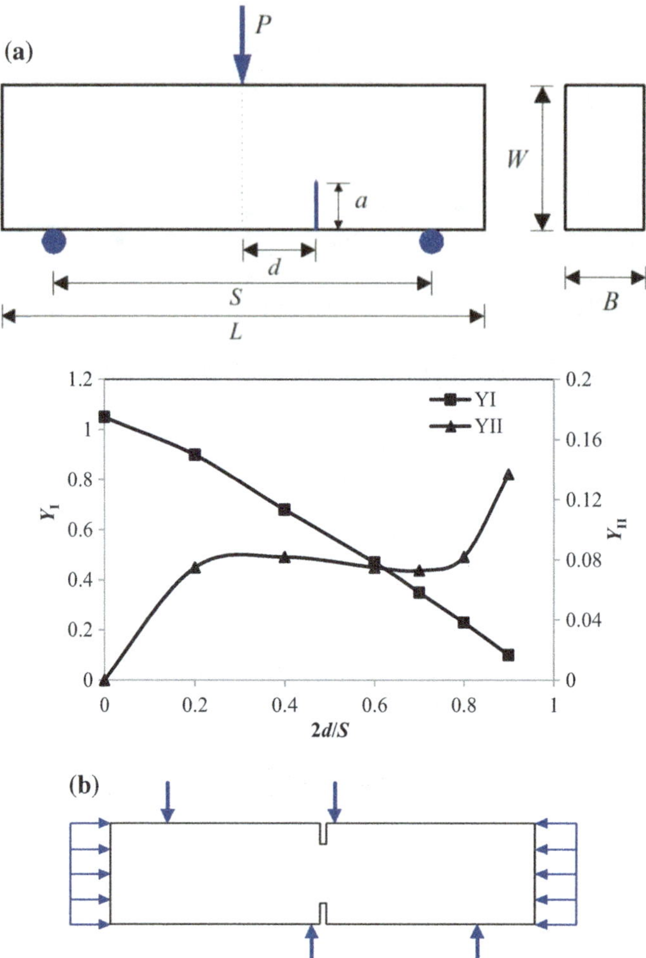

Fig. 3.2 **a** SENB specimen used for pure mode I and mixed mode I/II fracture tests on asphalt mixtures together with the values of geometry factors [5], **b** double-notched beam for performing pure mode II fracture test on asphalt mixtures [20]

where $M^e = 1$ corresponds to the pure mode I loading, $M^e = 0$ corresponds to the pure mode II loading, and any value of M^e between 0 and 1 (*i.e.*, $0 < M^e < 1$) refers to mixed mode I/II loading.

Pirmohammad and Ayatollahi [30] investigated the SCB specimen shown in Fig. 3.4 using finite element analysis to find the values of crack location (*i.e.*, L) and positions of the bottom supports (*i.e.*, S_1 and S_2) for generating different modes of loading. For this purpose, the radius and thickness of the specimen were assumed to be 75 mm and 32 mm, respectively, and a crack of length $a = 20$ mm was regarded within the specimen. In addition, a load of $P = 1000$ N was applied on top of the

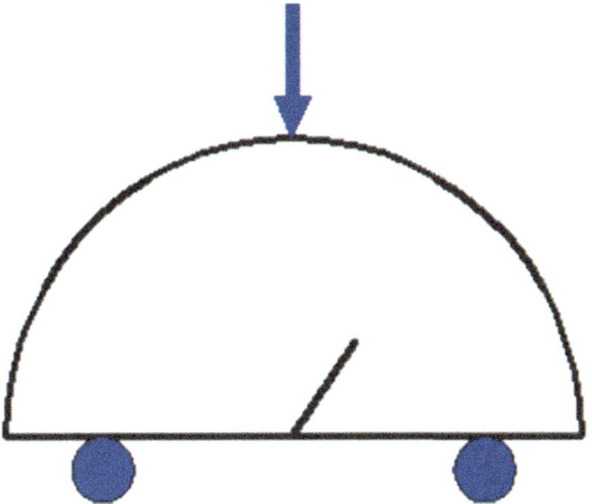

Fig. 3.3 SCB specimen used for performing mixed mode I/II fracture tests [5]

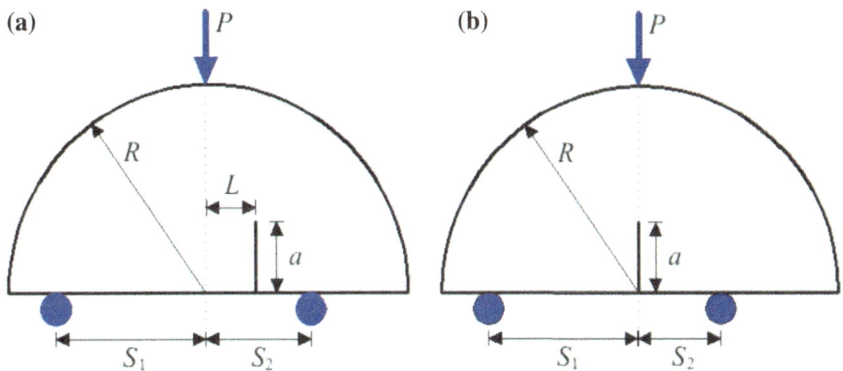

Fig. 3.4 SCB specimens for performing fracture tests under different mode mixities, **a** Type I, **b** Type II [29]

SCB specimen. It is also pointed out that the mechanical properties were regarded as: Young's modulus $E = 12.5$ GPa and Poisson's ratio $\upsilon = 0.35$. The SCB specimens were modeled using eight-node elements, and the J-integral method was chosen in the finite element analyses. In addition, singular elements with nodes at quarter-point positions were employed at the regions around the crack tip.

Table 3.1 presents the results of K_I and K_{II} directly extracted from the finite element analyses. According to Table 3.1, pure mode I (*i.e.*, $M^e = 1$) loading was achieved as the specimen was loaded symmetrically (*i.e.*, $S_1 = S_2 = 50$ mm), and the crack was located at the middle of the specimen (*i.e.*, $L = 0$). While, mixed mode I/II ($0 < M^e < 1$) and pure mode II ($M^e = 0$) loadings were achieved by asymmetrically loading of the specimen (*i.e.*, $S_1 = 50$ mm and $S_2 = 20$ mm), and as the crack was

Table 3.1 Numerical results for the SCB specimen shown in Fig. 3.4a [30]

Loading mode	M^e	$S_1, S_2,$ and L (mm)	K_I (MPa$\sqrt{m}$)	K_{II} (MPa$\sqrt{m}$)	Y_I	Y_{II}
Pure mode I	1	$(S_1, S_2) = (50, 50)$ and $L = 0$	0.195	0	3.734	0
Mixed mode I/II	0.8	$(S_1, S_2) = (50, 20)$ and $L = -2$	0.086	0.029	1.655	0.546
Mixed mode I/II	0.5	$(S_1, S_2) = (50, 20)$ and $L = 5$	0.061	0.059	1.171	1.131
Mixed mode I/II	0.2	$(S_1, S_2) = (50, 20)$ and $L = 11$	0.031	0.094	0.599	1.792
Pure mode II	0	$(S_1, S_2) = (50, 20)$ and $L = 16$	0	0.120	0	2.298

located at special position (see the values of L presented in Table 3.1). In other words, the proportion of mode II increased as the crack translated away from the middle of the specimen (or as the value of L enhanced). The stress intensity factors (*i.e.*, K_I and K_{II}) given in Table 3.1 were then replaced into the following relations to calculate the mode I and mode II geometry factors (*i.e.*, Y_I and Y_{II}):

$$Y_I = \frac{K_I}{\sqrt{\pi a}} \frac{2Rt}{P} \qquad (3.2)$$

$$Y_{II} = \frac{K_{II}}{\sqrt{\pi a}} \frac{2Rt}{P} \qquad (3.3)$$

The values of Y_I and Y_{II} are also given in Table 3.1. It is reminded that the geometry factors are required for calculating fracture toughness of asphalt mixtures.

3.2.3 Fracture Experiments

Usually, cylindrical samples made of asphalt mixtures are prepared using superpave gyratory compactor (SGC) in the laboratory. According to [30], these cylindrical samples were then sliced into several disks with special thicknesses using a water-cooled masonry sawing machine, and each disk was then halved into two semicircles. A crack was finally generated within the semicircular specimens using a water-cooled cutting machine with a very thin blade (more can be found in [31]). Figure 3.5 displays the cutting processes to produce SCB specimens.

In order to perform fracture experiments on the SCB specimens under different loading modes, the specimens were put upon the bottom supports adjusted at appropriate positions given in Table 3.1 (*i.e.*, $S_1 = S_2 = 50$ mm for pure mode I and $S_1 = 50$ mm and $S_2 = 20$ mm for mixed mode and pure mode II). The top fixture was

Fig. 3.5 Cutting processes for producing SCB specimen

then moved downward with a certain displacement rate. Load-*LLD* curve was finally recorded using a personal computer connected to the test apparatus (see Fig. 3.6).

The fracture resistance (or fracture toughness) of asphalt mixtures can be described by the values of critical stress intensity factors (critical SIFs) corresponding to the fracture load. The mode I and mode II critical SIFs (*i.e.*, K_{If} and K_{IIf}) can be calculated from the following relations:

$$K_{\mathrm{If}} = Y_{\mathrm{I}} \frac{P_{\mathrm{cr}}}{2Rt} \sqrt{\pi a} \tag{3.4}$$

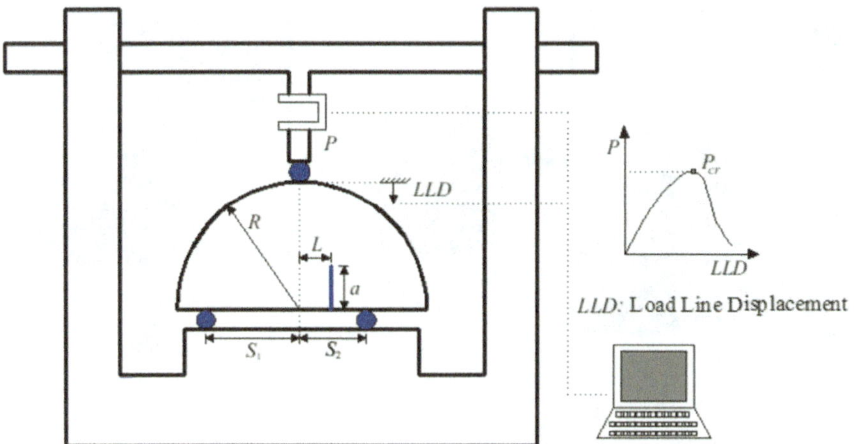

Fig. 3.6 Fracture test set-up

$$K_{\mathrm{IIf}} = Y_{\mathrm{II}} \frac{P_{\mathrm{cr}}}{2Rt} \sqrt{\pi a} \tag{3.5}$$

where the fracture load P_{cr} is the maximum load in the load-*LLD* curve (see Fig. 3.6). Y_{I} and Y_{II} are the mode I and mode II geometry factors. It is also reminded that R, t, and a are the specimen radius, the specimen thickness, and the crack length, respectively.

Fracture strength K_{eff} (called effective critical SIF) of materials subjected to any loading mode can be calculated from the following relation.

$$K_{\mathrm{eff}} = \sqrt{K_{\mathrm{If}}^2 + K_{\mathrm{IIf}}^2} \tag{3.6}$$

It is notable that the value of K_{eff} reduces to K_{If} and K_{IIf} for the cases of pure mode I and pure mode II, respectively.

It should be emphasized that Eqs. 3.4–3.6 are valid for the cases that fracture takes place in accordance with the linear elastic fracture mechanics (LEFM) approach, *i.e.*, when a brittle fracture occurs during fracture tests.

For the fracture tests under pure mode I loading using the SENB specimen shown in Fig. 3.2a, the fracture toughness can be calculated from the following equation [5]:

$$K_{\mathrm{If}} = Y_{\mathrm{I}} \frac{3S P_{\mathrm{cr}}}{2W^2 B} \sqrt{\pi a} \tag{3.7}$$

where

$$Y_{\mathrm{I}} = \frac{1.99 - \left(\frac{a}{W}\right)\left(1 - \frac{a}{W}\right)\left(2.15 - \frac{3.93a}{W} + \frac{2.7a^2}{W^2}\right)}{\sqrt{\pi}\left(1 + \frac{2a}{W}\right)\left(1 - \frac{a}{W}\right)^{3/2}} \tag{3.8}$$

Furthermore, for the mixed mode I/II tests using the SENB specimen shown in Fig. 3.2a, the following equations are used [5]:

$$K_{\mathrm{If}} = Y_{I}\frac{3SP_{\mathrm{cr}}}{2W^2B}\sqrt{\pi a} \tag{3.9}$$

$$K_{\mathrm{IIf}} = Y_{\mathrm{II}}\frac{3SP_{\mathrm{cr}}}{2W^2B}\sqrt{\pi a} \tag{3.10}$$

where Y_{I} and Y_{II} are the mode I and mode II geometry factors. Their values have been presented in Fig. 3.2a for the SENB specimen with the dimensions of: $L = 305$ mm, $W = 65$ mm, $B = 50$ mm, $S = 244$ mm, $a = 19.5$ mm, $a/W = 0.3$, $S/L = 0.8$.

It is reminded that the fracture strength K_{eff} of materials subjected to any loading mode can be calculated from the Eq. 3.6.

3.3 Effect of Aggregate Type

Braham et al. [32] conducted mode I fracture experiments using the DC(T) specimens on a broad range of material mixtures. They investigated the effect of aggregate type on fracture energy at three temperature levels (*i.e.*, -30, -18, and -6 °C). For this purpose, they prepared two mixtures containing granite andlimestone aggregates, while all other properties, for example, air void, binder type, and binder content, were identical for each mixture. According to the results, the granite mixtures showed fracture energy of 400–1100 J/m^2, while the limestone ones showed fracture energy of 200–700 J/m^2. Accordingly, the granite mixtures demonstrated higher fracture energy than the limestone mixtures at the lowest temperature (*i.e.*, -30 °C). The mastic had high strength at -30 °C; therefore, the overall fracture behavior of the mixture was more brittle. Conversely, at highest testing temperature (*i.e.*, -6 °C), the limestone and granite mixtures showed fracture energy of 2500–2700 J/m^2 and 2000–2500 J/m^2, respectively. Therefore, it can be concluded that at the warmer temperature, at which the crack propagated around the aggregates for both the granite and limestone mixtures, the mastic strength, morphology, or both, of the limestone mixtures imparted higher fracture strength. This can be because of the relatively higher physiochemical reactivity available in calcareous aggregates and fines, such as that available in the limestone mixture, as compared with that present in the more siliceous granite mixture [32–34]. In summary, the granite aggregates performed better than limestone ones at low temperatures, whereas this trend was reversed at high temperatures. Based on the fracture surfaces obtained from the DC(T) tests, the granite aggregates were not fractured at either low or high temperatures, while the limestone aggregates are only fractured at low temperatures, whereas at high temperatures, they remained intact.

In another study, granite and limestone aggregates were used by Li and Marasteanu [35] to investigate the effect of aggregate type on fracture energy using the SCB tests. In this study, a binder content of 6% for granite mixtures and 6.9% for limestone mixtures was used in the mix design. The SCB specimens used in this study were 150 mm in diameter and 25 mm in thickness. In addition, a crack of length 15 mm was carved within each specimen. The SCB tests were performed with a span of 120 mm at three different temperatures (*i.e.,* −30, −18, and −6 °C). A constant *CMOD* rate of 0.0005 mm/s was used in the SCB test, and the load-*LLD* curve was recorded to calculate the fracture energy (more details on the SCB test can be found in Chap. 5).

According to their results, the granite mixtures demonstrated significantly higher fracture energy than the limestone mixtures for all the testing temperatures. Visual inspection of the fracture surfaces of both mixtures showed that no obvious differences were observed between the fracture surfaces at −6 °C, *i.e.,* the fracture took place either in mastic or at the interface, and no aggregates fractured. However, at the two lower temperatures, a significant portion of the fracture took place through the aggregates made with limestone, while for the mixtures made with the granite, the crack traveled around the aggregates (see Fig. 3.7).

Li et al. [36] performed experiments on the granite and limestone mixtures to obtain the fracture energy and fracture toughness at three different temperature conditions (*i.e.,* −30, −18, and −6 °C) using the SCB specimen. Both the granite and limestone mixtures contained the same binder (*i.e.,* PG64-28) and air void percentage (*i.e.,* 4%). Their results revealed that for both the granite and limestone mixtures, the fracture energy reduced as the test temperature decreased, while fracture toughness increased when the test temperature reduced. The mixtures prepared with granite

(a)   **(b)**

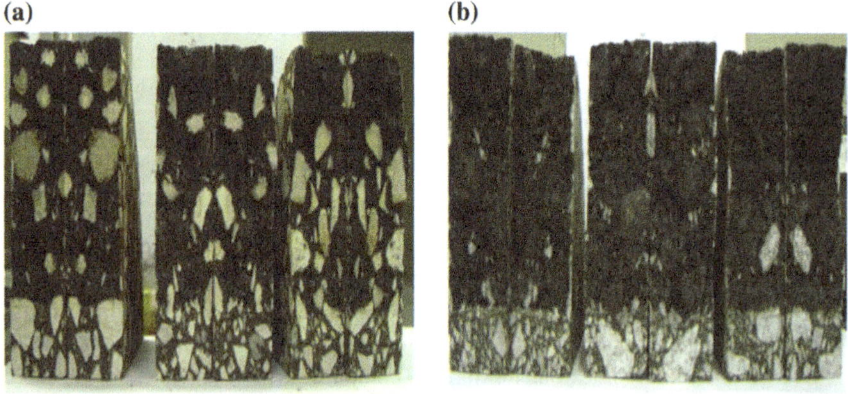

Fig. 3.7 Fracture surface of **a** limestone and **b** granite at −18 °C [35]

aggregates showed higher fracture energy and fracture toughness than those pre-pared with limestone aggregates at all the testing temperatures, indicating that it took more energy to fracture granite mixtures compared to limestone ones.

In another study, Aliha et al. [12] investigated the effect of aggregate type on fracture strength of asphalt concretes under different loading modes of I, II and mixed mode I/II. They used SCB specimens with 150 mm in diameter and 32 mm in thickness to calculate the fracture toughness of asphalt concretes containing limestone and siliceous aggregates. In addition, a crack with 20 mm in length was generated within each specimen. According to their results, asphalt concretes containing limestone aggregates had higher fracture strength than those containing siliceous aggregates for all loading modes due to the greater strength and stiffness of limestone aggregates relative to siliceous aggregates. Furthermore, the limestone mixtures performed better under pure modes of I and II than siliceous mixtures in comparison with the mixed mode I/II loading conditions. Meanwhile, limestone mixtures showed the greatest fracture toughness under pure mode II loading.

3.4 Effect of Aggregate Gradation

Pirmohammad and Ayatollahi [37] investigated the effect of aggregate gradation on the fracture resistance of asphalt mixtures. In this study, five different aggregate gradations designated by numbers 1, 3, 4, 5, and 6 were used, as shown in Table 3.2. According to Table 3.2, the NMAS for these aggregate gradations was 37.5 mm,

Table 3.2 Aggregate gradations used in the mixtures [37]

Sieve size (mm)	Passing percent (%)				
	A.G. No.[a] 1	A.G. No. 3	A.G. No. 4	A.G. No. 5	A.G. No. 6
50	100	–	–	–	–
37.5	95	–	–	–	–
25	84	100	–	–	–
19	68	95	100	–	–
12.5	58	81.5	95	100	–
9	48	68	77	95	100
4.75	38	50	59	70	90
2.36	28	36	43	49.5	83
1.18	22	28	33	38	60
0.5	16	20	23	26.5	45
0.3	10	12	13	15	24
0.15	7	8.5	9.5	10.5	12
0.075	3	5	8.4	6	6

[a]A.G. No.: Aggregate Gradation Number

Fig. 3.8 Conventional three-point bent fracture test set-up

19 mm, 12.5 mm, 9 mm, and 4.75 mm, respectively. In the preparation of these mixtures, the same air void content (*i.e.,* 4%) and binder type (*i.e.,* penetration grade of 60/70) were used, *i.e.,* the aggregate gradation was the only parameter for investigation. As an exception, it is mentioned that the air void content was 8.5% for the HMA mixture containing aggregate gradation No. 1. Because it was not possible to compact the mixture further for achieving air void content of 4%, *i.e.,* by more compaction, the large aggregates in the HMA mixture would break, which is not recommended in the HMA preparation process.

In this study, fracture experiments were performed on the SCB specimens under different loading modes at −10 °C. Firstly, each specimen was put into a freezer fixed at −10 °C for two hours to ensure that the entire SCB specimens had an identical temperature. It is worth noting that two hours were found to be sufficient time for avoiding any temperature gradient within the asphalt concrete samples [35]. The fracture experiments were immediately conducted using the conventional three-point bend set-up shown in Fig. 3.8, in which a displacement-controlled system was utilized to load the SCB specimens. The two bottom supports were fixed, and the top fixture was moved downward by a constant speed of 3 mm/min. Four SCB specimens were tested for each mode of loading.

Figure 3.9a displays the mode I fracture resistance of the mixtures containing various aggregate gradations (*i.e.,* No. 1, No. 3, No. 4, No. 5, and No. 6). According to this figure, by decreasing the aggregate gradation number from 6 to 3, the value of K_{If} increased, indicating that the mixtures containing coarser aggregates had higher fracture resistance. For the HMA mixture containing the aggregate gradation No. 1, although its aggregates were the coarsest ones among the mixtures investigated in this study, its fracture resistance was the lowest one due to having higher air void content of 8.5%. Air voids in HMA mixtures act as stress concentrators that may lead to a reduction in the fracture resistance. Hence, this HMA mixture was not considered a suitable mixture, because in a mixture with good gradation, fin aggregates should

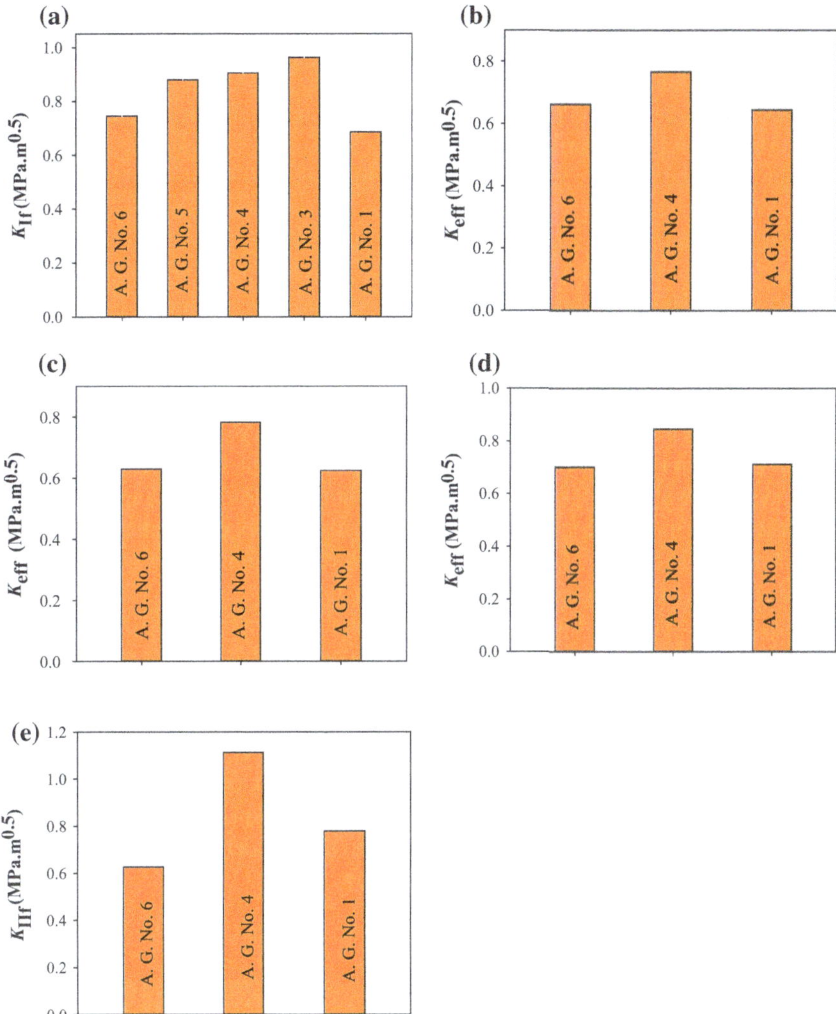

Fig. 3.9 Fracture resistance of HMA mixtures containing different aggregate gradations under **a** pure mode I, **b** mixed mode of $M^e = 0.8$, **c** mixed mode of $M^e = 0.5$, **d** mixed mode of $M^e = 0.2$, **e** pure mode II [37]

fill spaces between the coarse aggregates. Such mixtures are often used in the lower layers (for example, in the base layer) of the pavement systems.

Figure 3.9b–e exhibits the mixed mode I/II and pure mode II fracture resistances of HMA mixtures containing different gradation numbers. Similar to the results obtained for the mode I loading, the mixed mode I/II, and pure mode II fracture resistances of the mixture containing aggregate gradation No. 4 were more than those of the mixtures containing aggregate gradation No. 6 and No. 1. Some reasons have been suggested by researchers for improvement in the strength of mixtures when applying coarse aggregates. For example, according to Christensen et al. [38], the rutting resistance was improved by employing coarse aggregates because coarse aggregates in the mixture increased the internal friction in the mixture due to good interlocking between the aggregates. Furthermore, mixtures containing finer aggregates showed lower strength due to being more prone to micro-cracking [39].

Since brittle fracture was observed for the mixtures considered in this study, the fracture in the SCB specimens was controlled only by the binders and aggregates available around the crack front. Figure 3.10 illustrates fracture surface of the mixtures containing aggregate gradation numbers of 1, 4, and 6. By changing the aggregate gradation numbers from 6 to 1, the binder content of the mixtures reduced, whereas the amount of aggregates at the crack front increased. On the other hand, as is well known, the aggregate had higher tensile strength than the binder [16]. Hence, the mixtures having coarser aggregates were expected to show higher fracture toughness. However, the higher aggregate content, larger aggregates, and less binder content around the crack tip were responsible for improving the fracture resistance of the mixture containing aggregate gradation No. 4. It is also noticed that the higher air void content (*i.e.*, 8.5%) in the mixture containing aggregate gradation No. 1 decreased its mixed mode I/II fracture resistance.

Based on the results of fracture resistance given in Fig. 3.10 for the HMA mixtures containing various aggregate gradations, it can be deduced that by increasing the shear load (by decreasing the value of M^e), the fracture resistance initially reduced and then increased. For the mixtures containing aggregate gradation numbers of 4 and 1, the fracture resistance reached its maximum value under pure mode II loading, except for the mixture containing aggregate gradation No. 6 for which the mode II fracture resistance reduced remarkably. Hence, it can be concluded that the mixture containing finest aggregates had less fracture strength under pure mode II loading because of existing poor interlock between the aggregates.

(a)   **(b)**   **(c)**

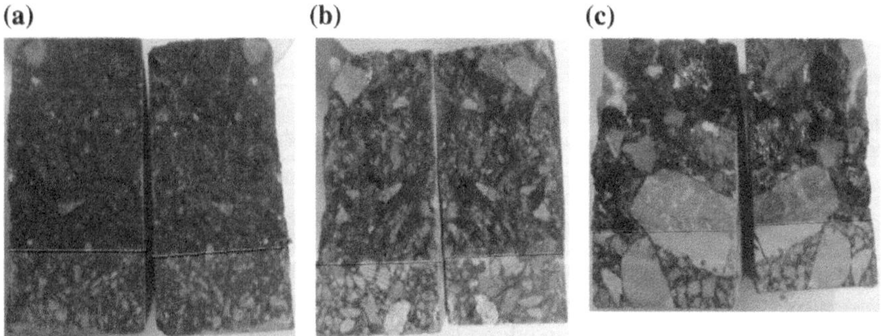

Fig. 3.10 Fracture surface of the HMA mixtures containing aggregate gradation **a** No. 6, **b** No. 4, **c** No. 1, under pure mode I [37]

Table 3.3 Asphalt mixture aggregate gradations [12]

Sieve size (mm)	Passing percent (%)		
	NMAS: 12.5 mm	NMAS: 9 mm	NMAS: 4.75 mm
19	100	100	100
12.5	95	100	100
9	80	85	100
4.75	59	70	90
2.36	43	49.5	82.5
1.18	30	33	60
0.5	18	22	45
0.3	13	15	23.5
0.15	8	10	11.5
0.075	6	6	10

Aliha et al. [12] conducted experiments on the SCB specimens under different loading modes to obtain fracture toughness of HMA mixtures made up of different aggregate gradations shown in Table 3.3. All the experiments were performed at -15 °C and a binder with penetration grade of 60/70 was used in the mixtures. Their results exhibited that the value of fracture toughness decreased as the size of aggregates became small. Mixtures containing finer aggregates demonstrated lower fracture strength under shear loading than that under tensile loading. Hence, coarser aggregates were more effective for improving the fracture strength of mixtures subjected to shear loads. As the size of aggregates in the mixture reduced, the risk of micro-crack initiation inside the wider space of filler and binder part of asphalt mixture increased, which made the mixture more vulnerable to failure due to application of shear loads. It is worth noting that similar results have been observed by Pirmohammad and Ayatollahi [37], as explained above.

3.5 Effect of Air Void Content

Pirmohammad and Ayatollahi [37] investigated the effect of air void content on the fracture resistance of HMA mixtures. They conducted several experiments using the SCB specimens under different types of loading: pure mode I, pure mode II, and mixed mode I/II at -10 °C. Three HMA mixtures with different air void contents of 4, 7, and 10% were prepared in this study. All the mixtures contained aggregate gradation number of 4 presented in Table 3.2 and binder with penetration grade of 60/70.

Figure 3.11 shows fracture resistance of the HMA mixtures containing various air void percentages. As seen in this figure, the fracture resistance decreased as the air void content increased. Indeed, percentage of the air void influenced the mixture

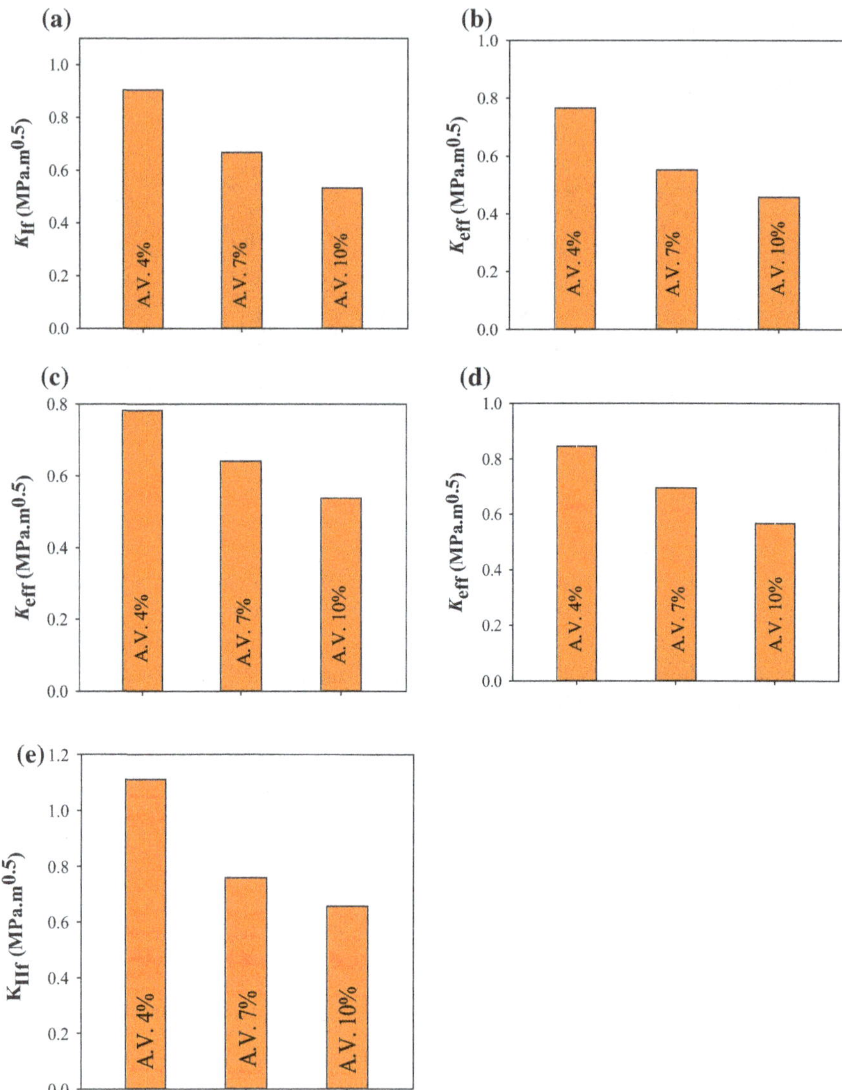

Fig. 3.11 Fracture resistance of HMA mixtures containing different air void contents under **a** pure mode I, **b** mixed mode ($M^e = 0.8$), **c** mixed mode ($M^e = 0.5$), **d** mixed mode ($M^e = 0.2$), **e** pure mode II [37]

density. More compacted mixture has higher density, and a dense mixture requires more load and strain energy for crack extension. In addition, air voids in the mixture were sources of stress concentration; hence, they can accelerate the crack extension at low temperatures.

Similar results have been reported by Aliha et al. [12]. They investigated the effect of air void on the fracture toughness of HMA mixtures containing three different air void contents of 3, 5, and 7%. Based on the results, the fracture toughness reduced as the air void content in the mixture increased.

According to Li and Marasteanu [35], the mixtures with 4% air void resulted in higher fracture energy than those with 7% air void content, as expected. Because more strain energy was required to break a more compacted asphalt mixture. It is also pointed out that the presence of air void decreases the overall mixture strength, but overall fracture process zone behavior may be affected in a complex manner, since air voids may also act to blunt a sharp crack tip [40].

3.6 Effect of Binder Content and Binder Type

In an investigation conducted by Braham et al. [32], the effect of binder content was studied at three temperature levels (*i.e.,* -30, -18, and -6 °C). Two different asphalt contents were used in this study: (i) the first binder content was the optimal binder content found from the superpave mix design procedure, (ii) the second binder content was to increase the optimal binder content by 0.5%. In this study, all other properties including air void, binder type, and aggregate type were the same for each mixture, *i.e.,* the binder content was the only difference. Based on the results, at the low and mid temperatures, the addition of binder did not affect the fracture energy due to creating a brittle matrix at low temperatures, whereas at high temperatures, the addition of 0.5% binder increased the fracture energy significantly because the binder can dissipate tremendous amounts of energy. Hence, more binder content does not necessarily increase the fracture energy at low temperatures but does increase the fracture energy at high temperatures.

Mobasher et al. [1] investigated the effect of binder content on mode I fracture toughness of HMA mixtures at two test temperatures of -1 and -7 °C. They prepared HMA mixtures with binder contents of 4, 5 and 6% for performing the experiments on cracked SENB specimens. According to their results, the mixtures with 5% binder content showed the highest value of fracture toughness at both the test temperatures of -1 and -7 °C.

Li and Marasteanu [35] investigated the effect of the binder modifier (*i.e.,* SBS) and binder type on the fracture energy using the SCB tests at -30, -18, and -6 °C. According to the results, the asphalt mixtures modified with SBS had obviously higher fracture energy than those containing plain binder. This improvement in fracture energy was more pronounced at the two lower temperatures, such that the modified mixtures showed more than 30% increase of fracture energy compared to the unmodified mixture. It is also shown that the asphalt mixture containing PG58-28 binder had higher fracture energy than that containing PG64-28 binder, since the PG58-28 binder is softer than the PG64-28 binder. Hence, asphalt mixture containing PG58-28 binder was more resistant to fracture than that containing PG64-28 binder.

Pirmohammad and Ayatollahi [37] investigated the effect of binder type on the fracture strength of HMA mixtures using the SCB specimen. The tests on the SCB specimens were conducted under different loading modes (*i.e.*, pure mode I, pure mode II, and mixed mode I/II). Four HMA mixtures were prepared in this study, which three of them contained binders with penetration grades of 40, 60, and 85, and the fourth mixture contained a binder with penetration grade of 85 modified with 3.5% by weight of SBS. All these mixtures contained aggregate gradation number of 4 presented in Table 3.2 and air void content of 4%. The experiments were conducted at $-10\,^\circ$C.

Figure 3.12a shows the mode I fracture strength (K_{If}) for the mixtures containing various binders, in which the fracture strength increased as the penetration grade of the binder enhanced. Furthermore, more improvement in fracture strength was observed by applying a little amount of SBS in the mixture. Indeed, the penetration grade of a binder states its softness; hence, by increasing the penetration grade of a binder, softness of the mixture increases. This increase in the softness was responsible for increasing the fracture strength of HMA mixtures at the test temperature of $-10\,^\circ$C, at which HMA mixtures often show brittle behavior. It is pointed out that in brittle materials, the formation of micro-cracks around the crack front results in premature fracture. Therefore, employing binders with higher penetration grade can postpone the process of crack growth by local plastic deformations around the crack front. Meanwhile, in addition to the softening effect of SBS, it also improves the adhesion between the aggregate and binder [41–43]. However, application of inappropriate amount of SBS would worsen the fracture strength [44]. It is also notable that softer binders can subside the brittleness of asphalt concretes resulting from the aging effects [45].

Figure 3.12b–e represents the mixed mode I/II and pure mode II fracture strengths of the mixtures containing various binder types. Similar to the results observed for the mode I loading, the mixed mode I/II and pure mode II fracture strengths increased as the penetration grade increased.

In another study, Aliha et al. [12] investigated the effect of binder type on the fracture toughness by preparing mixtures containing binders with penetration grades of 60/70 and 85/100. According to their results, the soft binder (*i.e.*, 85/100) improved fracture toughness of asphalt mixtures significantly. Particularly, this binder type more increased the pure mode II fracture toughness than that of the pure mode I.

3.7 Effect of Temperature

Many studies can be found in the literature investigating the effect of temperature on fracture properties of asphalt mixtures. Reviewing the literature indicates that the researchers have characterized two important fracture indicators including the fracture toughness and fracture energy to consider the effect of temperature. Some of the important researches are discussed herein.

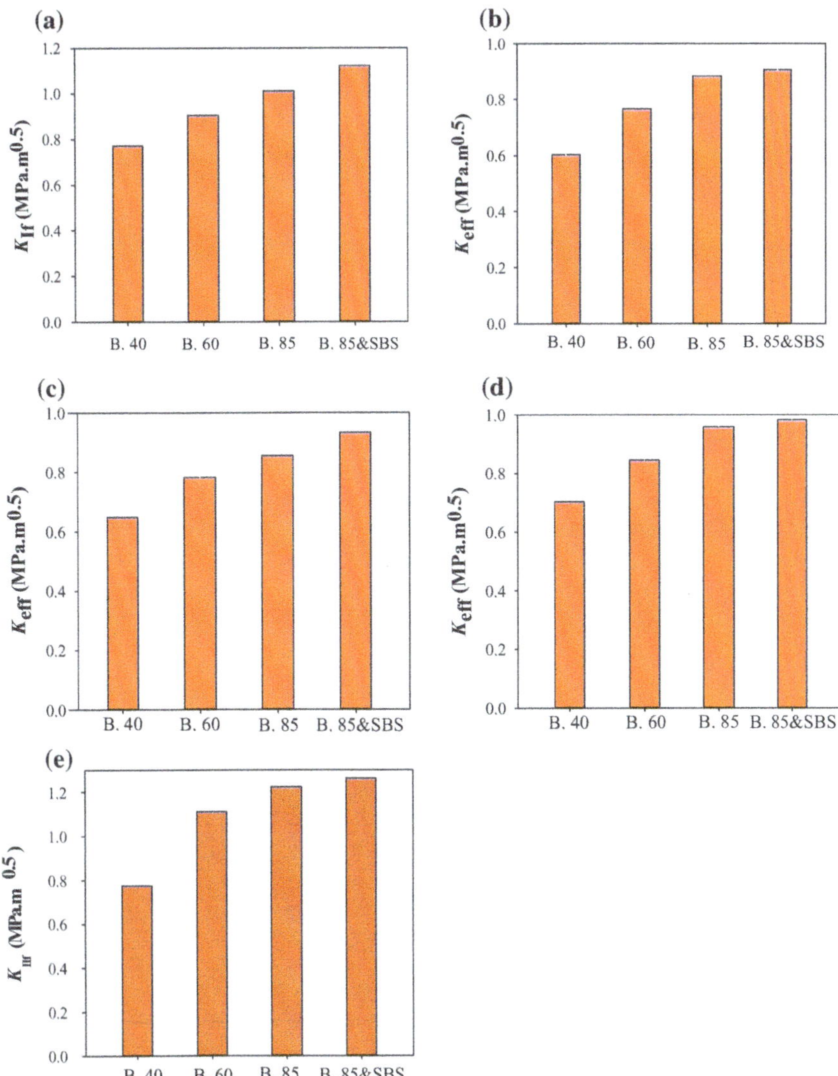

Fig. 3.12 Fracture resistance of HMA mixtures containing various binder types under **a** pure mode I, **b** mixed mode (*i.e.*, $M^e = 0.8$), **c** mixed mode ($M^e = 0.5$), **d** mixed mode ($M^e = 0.2$), **e** pure mode II [37]

Braham et al. [32] have investigated the effect of temperature (*i.e.*, -6, -18 and -30 °C) on fracture energy by performing experiments on the DC(T) specimens under pure mode I loading. The results indicated that the fracture energy decreased as the test temperature reduced from -6 to -30 °C. According to the results, at low temperatures, the fracture energy was small due to the brittle type of fracture in asphalt mixtures, while the fracture energy significantly increased at high temperatures because the fracture path was observed to travel around the aggregates which dissipated more energy.

Kim and Hussein [3] investigated the effect of temperature on the mode I fracture resistance of asphalt concretes using the SENB specimen. In this study, a binder with penetration grade of 85/100 and limestone aggregates with maximum size of 12.7 mm were used for making asphalt concretes. The SENB specimens had the length, width, thickness, notch length, and span length of 300 mm, 70 mm, 50 mm, 15 mm, and 280 mm, respectively.

The SENB specimens were aged for one day at the ambient temperature before putting into the cold chamber. The specimens were then conditioned at the specified temperatures (*i.e.,* −5, −10, −20, −25, and −30 °C) for three days. Afterward, the SENB specimens were loaded using a three-point bend test set-up. The load was vertically applied to the specimens at the rate of 3.24 mm/min. Figure 3.13 plots the results of fracture toughness. As shown in this figure, the mode I fracture toughness of the asphalt concrete increased by dropping the temperature from −5 °C to −15 °C and then decreased thereunder.

Kim et al. [4] also investigated the effect of temperature on different asphalt concretes using the SENB specimen. Binder with penetration grade of 85/100 was used as a base asphalt binder. The asphalt concrete made of this binder is designated as AP, and the asphalt concretes modified by a low-density polyethylene (LDPE), a styrene-butadiene-styrene (SBS), and a mixed polymer of LDPE and SBS are designated as LP, SB, and LS, respectively. Aggregates with a maximum size of 19 mm were used in this research. The mixtures also had an air void content of 4%.

The SENB specimens had the length, width, thickness, notch length, and span length of 340 mm, 75 mm, 55 mm, 15 mm, and 300 mm, respectively. Three specimens were produced for each material and each test temperature, and the average results were reported.

The SENB specimens were aged for one day at the room temperature before putting into the environmental chamber. The specimens were then conditioned at

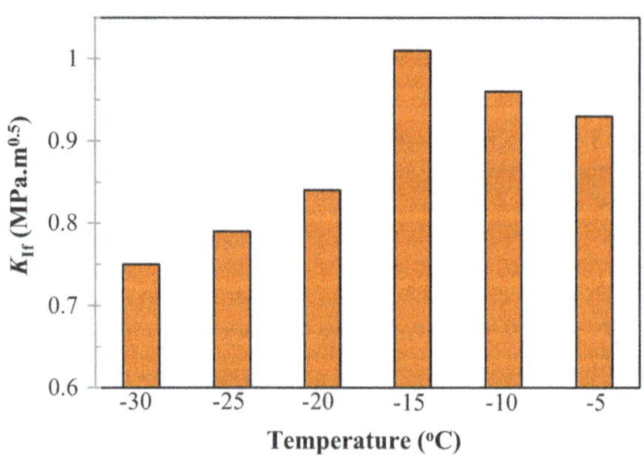

Fig. 3.13 Mode I fracture toughness of limestone asphalt concrete at different temperatures [3]

the specified temperatures (*i.e.,* -5, -10, -20, -30, and -35 °C) for two days. Afterward, the SENB specimens were loaded using a three-point bend test set-up. The load was vertically applied to the specimens at the rate of 3.35 mm/min.

Based on the results, the peak value of mode I fracture toughness was achieved at -5, -30, -30, and -10 °C for the mixtures designated as AP, LP, SB, and LS, respectively.

Artamendi and Khalid [5] measured the mode I fracture energy of two types of asphalt mixtures [*i.e.,* dense bitumen macadam (DBM) and stone mastic asphalt (SMA)] at the temperatures of $-10, 0,$ and 10 °C using the SCB and SENB specimens. Figure 3.14 plots the results of fracture energy. According to this figure, generally, the fracture energy increased as the temperature enhanced. Details of this research have been given in Chap. 5, Sect. 5.2.3.

In another study, Braham and Buttlar [20] characterized the mode II fracture energy of asphalt concrete at three temperatures (*i.e.,* -6, -18, and -30 °C) using the double-notched beam specimen shown in Fig. 3.2b. The beam had a length, width, and thickness of 165 mm, 50 mm, and 75 mm, respectively. In addition, the notch length on the top and bottom of the beam was 8 mm, producing a ligament length of 34 mm. The beams were prepared from asphalt concretes with a 12.5 mm NMAS and a 120/150 binder.

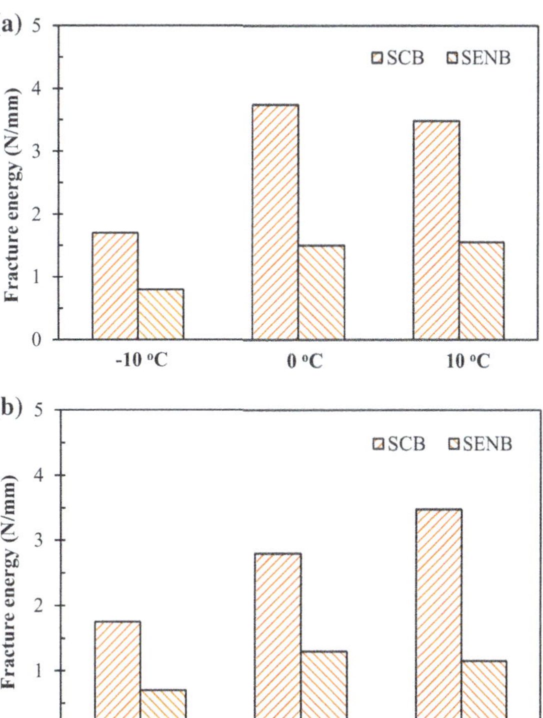

Fig. 3.14 Fracture energy for the **a** SMA and **b** DBM mixtures obtained from the SENB and SCB tests at different temperatures [5]

Tests were performed under controlling *LLD*. The *LLD* was run at a constant rate of 5 mm/min. Their results exhibited that the mode II fracture energy of asphalt concrete increased as the test temperature enhanced due to reducing the brittleness of asphalt concrete.

Pirmohammad and Ayatollahi [30] investigated the effect of temperature on the fracture resistance of HMA mixtures using the SCB specimen shown in Fig. 3.4a. The HMA mixtures considered in this research contained aggregates with gradation No. 4 (presented in Table 3.2), binder with penetration grade of 60/70, and air void of 4%. In addition, the SCB specimens used in this research had a diameter of 150 mm, thickness of 32 mm, and crack length of 20 mm. It is also reminded that the crack location in the SCB specimen for simulating different modes of loading together with the corresponding values of the geometry factors is given in Table 3.1.

The experiments were performed under different types of loading: pure mode I, pure mode II, and mixed mode I/II at four temperature levels namely 0, -10, -20, and -35 °C. For this purpose, the SCB specimens were initially cooled in a freezer set at the test temperature for 12 h to ensure that the entire SCB specimens have an identical temperature. The fracture experiments were then performed using the three-point bend set-up as shown in Fig. 3.8 at a constant displacement rate of 3 mm/min. Four replicates were tested for each temperature to reduce the discrepancies in the results.

Figure 3.15 displays the effect of temperature on the fracture strength of asphalt concrete, in which for all the modes of loading, the fracture strength initially increased to reach its peak value at -20 °C and then decreased as the test temperature reduced. Increase in the fracture strength of asphalt concrete for the temperatures above -20 °C is imputed to the increase in the strength of binder due to its contraction. In other words, more energy was required to break the binder as the temperature reduced. Furthermore, the adhesion between aggregates and binders enhanced as the temperature reduced [4]. On the other hand, decrease in the fracture toughness of asphalt concrete below -20 °C can be attributed to appearance of an internal damage known as differential thermal contraction (DTC). As the temperature drops, both aggregate and binder contract, but the binder surrounding the aggregate particle contracts more than the aggregate because of the significant difference in the coefficients of thermal expansion of aggregate and binder. Hence, the tensile stresses developed within the binder lead to the formation of hairline (micro) cracks which results in weakening the asphalt concrete (see Fig. 3.16). At relatively high temperatures (*i.e.,* above 0 °C), binder may flow and relax such stresses, but binder, and therefore, asphalt concrete has a brittle behavior at low temperatures, and the mentioned tensile stresses may build up. Consequently, fracture occurs within the binder as the tensile stress exceeds its strength [46, 47].

In another study, Mansourian et al. [48] investigated the effect of temperature on the plain and modified warm mix asphalt concretes. The plain mixture contained binder with penetration grade of 85/100 and 3% of Sasobit (by weight of the binder). The modified warm mix asphalt concrete contained jute fibers with 20 mm in length and three different dosages (*i.e.,* 0.3, 0.5, and 0.7%, by weight of the mixture). They performed the fracture tests at three temperature conditions (*i.e.,* 0, -10, and -20 °C)

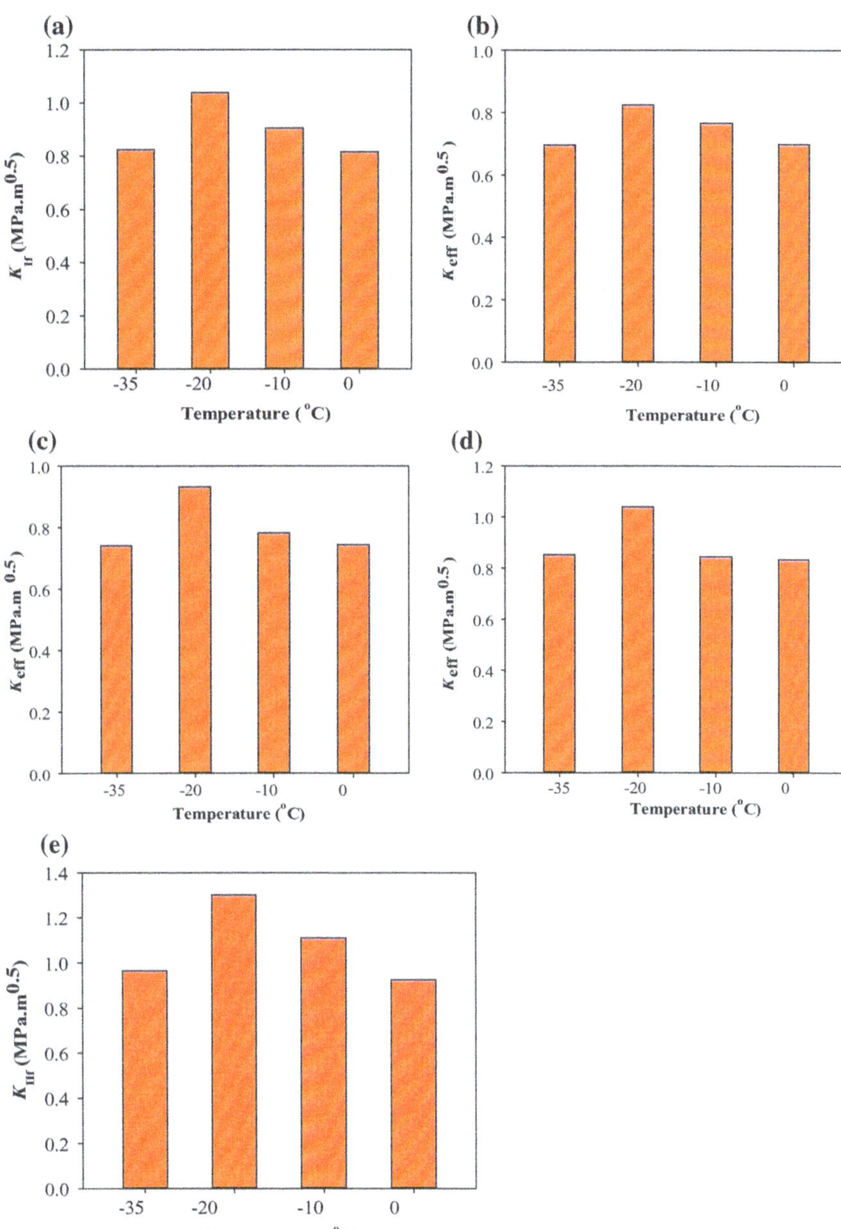

Fig. 3.15 Fracture resistance of HMA mixtures at different temperatures under **a** pure mode I, **b** mixed mode of $M^e = 0.8$, **c** mixed mode of $M^e = 0.5$, **d** mixed mode of $M^e = 0.2$, **e** pure mode II [30]

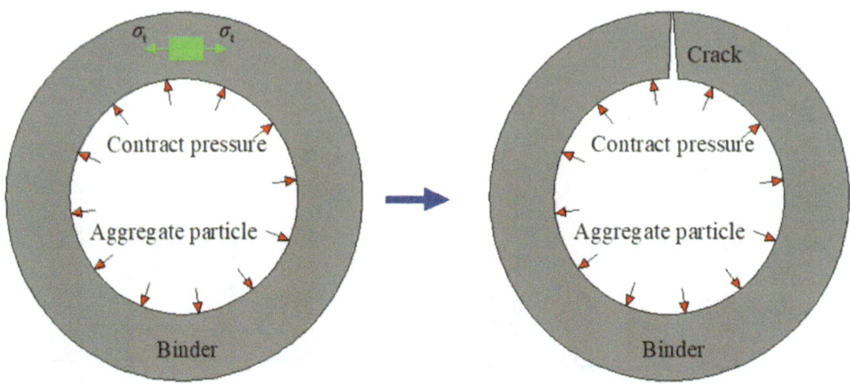

Fig. 3.16 DTC damage occurred in the binder [4]

under different modes of loading (*i.e.*, pure mode I, pure mode II, and mixed modes of $M^e = 0.2$ and 0.5) using the SCB specimen shown in Fig. 3.4a. Their results showed that for all the modes of loading, the fracture resistance of both the plain and modified warm mix asphalt concretes increased as the temperature dropped.

Pirmohammad and Kiani [31, 49] investigated the effect of temperature variations on the fracture strength of asphalt concrete under different modes of loading. The SCB specimens used in this research had a diameter of 150 mm, thickness of 32 mm, and crack length of 20 mm. The HMA mixtures contained aggregates with gradation No. 4 (presented in Table 3.2), binder with penetration grade of 60/70, and air void of 4%. It is also reminded that the crack location in the SCB specimen for simulating different modes of loading together with the corresponding values of the geometry factors is given in Table 3.1.

Fracture experiments were performed at the following climate conditions:

(i)   The specimens were conditioned at a constant temperature (CT) of $-15\,°C$ for 4 h;

(ii) The specimens were exposed to a variable temperature (VT) condition. Figure 3.17 plots profile of the VT condition, in which the temperature decreases from 25 to $-30\,°C$ and then increases to 25 °C. It is notable that the rate of temperature change was 6 h per half cycle. Meanwhile, the specimens experienced the VT condition for seven days before performing the tests.

In order to compare the results of the VT condition with those of the CT one, the specimens conditioned at the VT were finally kept at the temperature of $-15\,°C$ for 4 h similar to the CT condition.

After exposing the specimens to the abovementioned temperature conditions (*i.e.*, CT or VT), they were put upon the bottom supports of the universal test machine (UTM) adjusted at the proper locations (*i.e.*, S_1 and S_2 given in Table 3.1), and the top fixture was finally moved downward with a displacement rate of 3 mm/min to load the specimens (see the test set-up in Fig. 3.6). It is pointed out that the experiments

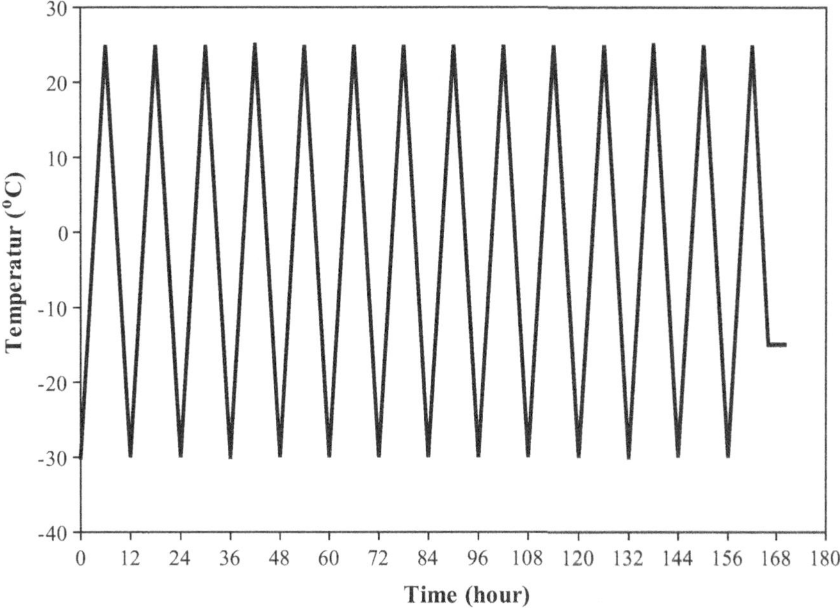

Fig. 3.17 Temperature profile for the VT condition [31]

were conducted under pure mode I, pure mode II, and three different mixed modes of $M^e = 0.8$, $M^e = 0.5$, and $M^e = 0.2$.

Figure 3.18 displays the SCB specimens after fracture tests under different modes of loading. For the pure mode I loading, the crack initiated along the initial crack line and developed straightly toward the top fixture while for the mixed mode I/II and pure mode II loadings, the crack kinked from the initial crack line and propagated along a curvilinear path. This manner of the crack growth path is attributed to the maximum tensile stress around the crack front, which was not anymore along the initial crack as the SCB specimen was loaded under mixed mode I/II or pure mode II [50]. Meanwhile, as the proportion of shear mode (*i.e.*, mode II) at the crack front of the SCB specimen increased (*i.e.*, as M^e decreases), the initiation angle of fracture became greater.

Figure 3.19 compares the fracture strength of asphalt concretes exposed to the climate conditions of CT and VT under various modes of loading. According to this figure, the fracture toughness of the specimens conditioned at the VT was less than the CT under any mode of loading, indicating that the climate condition of VT worsened the crack growth resistance of asphalt pavements. This may be attributed to the internal damage (known as DTC) occurred in the mixtures due to existing considerable difference in coefficients of thermal expansion of aggregate and binder. As mentioned earlier, by decreasing temperature, the fracture toughness of asphalt concretes initially increased to reach its peak value at a certain temperature and then decreased. The certain value of temperature, at which the trend of fracture toughness

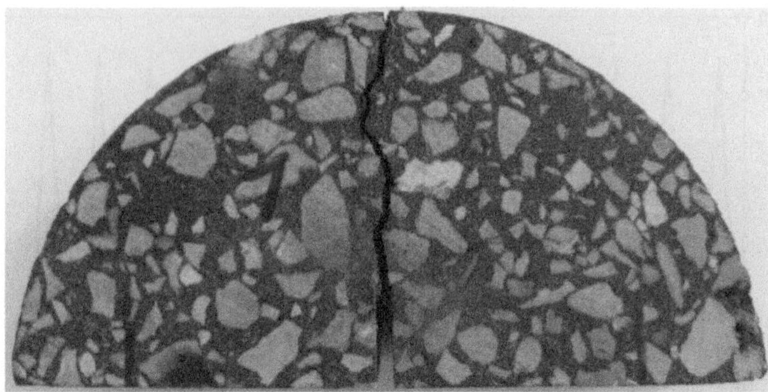

Pure mode I (M^e=1)

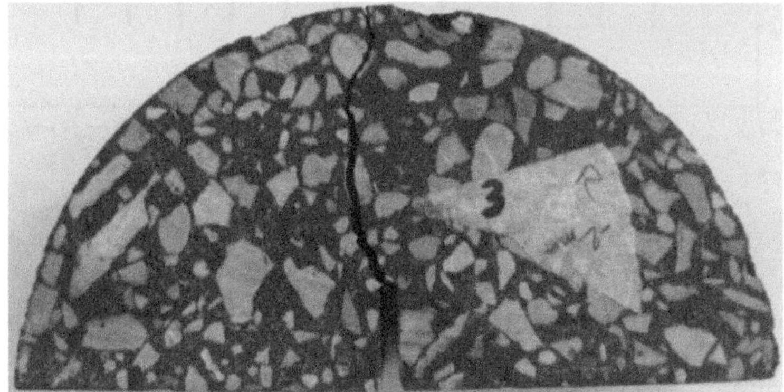

Mixed mode I/II (M^e=0.8)

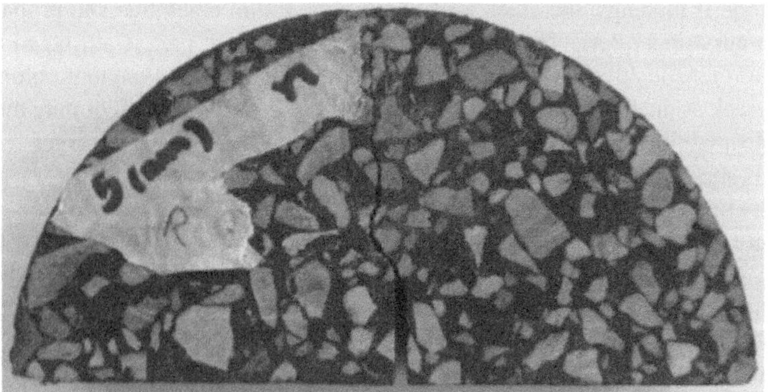

Mixed mode I/II (M^e=0.5)

Fig. 3.18 The SCB specimens after fracture tests under different modes of loading [31]

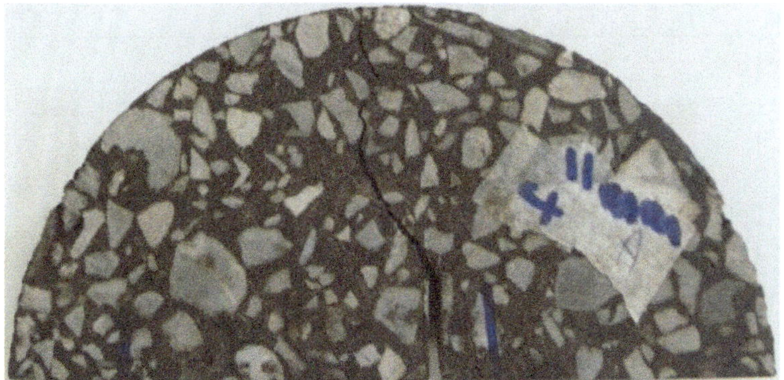

Mixed mode I/II (M^f=0.2)

Pure mode II (M^f=0)

Fig. 3.18 (continued)

reverses, depends on the composition of asphalt concrete, particularly the binder type. For example, its value was -15 and $-20\,°C$ for the mixtures used by Kim and Hussein [3] and Pirmohammad and Ayatollahi [30], respectively. Considering these explanations, as the HMA mixtures studied in this research were exposed to the climate condition of VT, they experienced the temperatures below $-20\,°C$ (see Fig. 3.17), and thereby, it was vastly likely to form micro-cracks within the binder surrounding the aggregate due to the DTC damage. Furthermore, generation of the micro-cracks increased as the mixtures experienced further cycles of temperature variation. Consequently, fracture strength of the HMA mixtures conditioned at VT was expected to decrease.

Based on the results given in Fig. 3.19, the asphalt concrete showed its minimum value of fracture toughness under the mixed mode of $M^e = 0.5$ for both the CT and VT climate conditions.

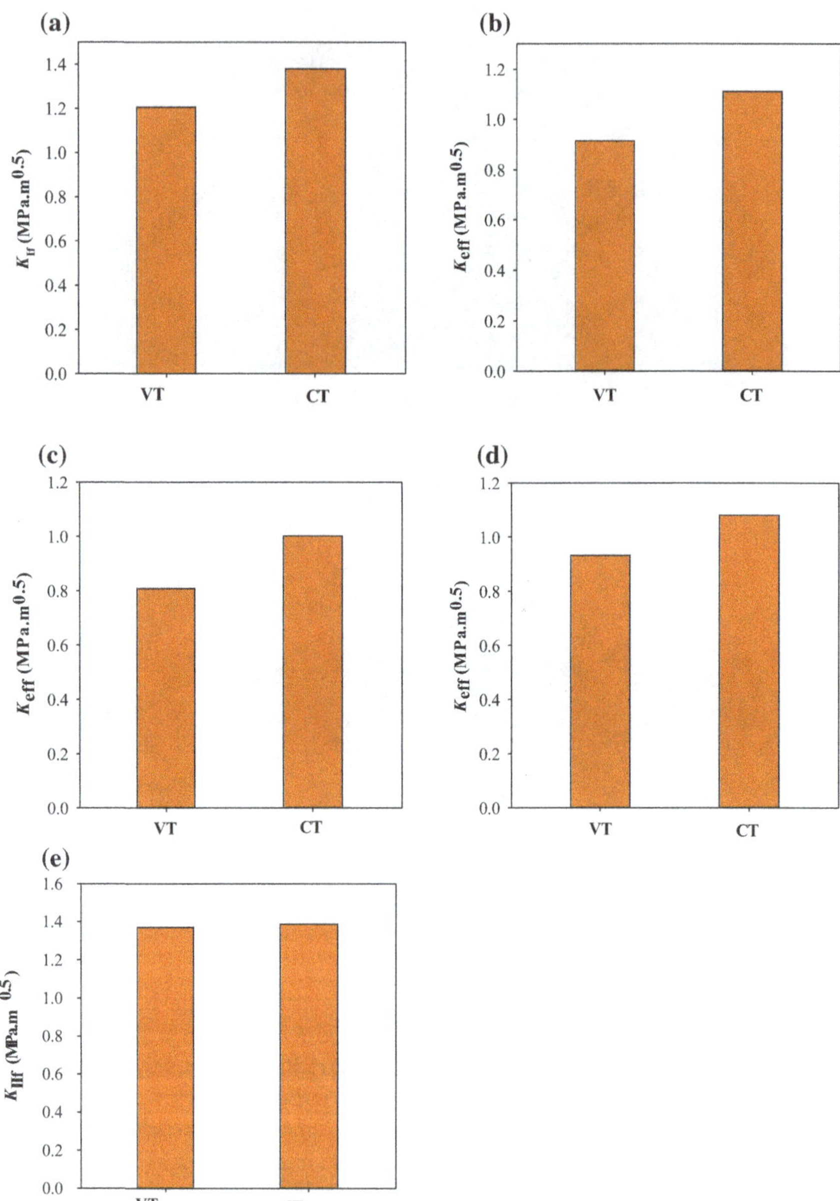

Fig. 3.19 Fracture strength of asphalt concretes exposed to the climate conditions of CT and VT under **a** pure mode I, **b** mixed mode (*i.e.*, $M^{e} = 0.8$), **c** mixed mode ($M^{e} = 0.5$), **d** mixed mode ($M^{e} = 0.2$), **e** pure mode II [31]

In another study, Pirmohammad and Shabani [51] investigated the effect of temperature variations on the fracture resistance of HMA concretes modified by SBS and CR (crumb rubber). The HMA mixtures considered in this research contained aggregates with gradation No. 4, binder with penetration grade of 60/70 (as a base binder), and air void of 4%. The SBS modified HMA mixture and CR modified one were respectively prepared by 5% SBS and 15% CR by weight of the base binder.

In order to prepare the SBS modified binder, powders of SBS were mixed with the base binder at 177 °C for 2 h. As such, for preparing the CR modified binder, particles of the CR were added to the base binder at 170 °C and were then mixed for 240 min using a high shear homogenizer mixer at 5500 rpm.

The SCB specimens used in this research had a diameter of 150 mm, thickness of 32 mm, and crack length of 20 mm. The supports and crack locations in the SCB tests for simulating different modes of loading have been given in Table 3.1.

The fracture experiments were conducted under pure mode I, pure mode II, and three different mixed modes of $M^e = 0.8$, $M^e = 0.5$, and $M^e = 0.2$. In addition, the tests were conducted at two different climate conditions of CT (at -15 °C) and VT (according to Fig. 3.17) similar to the conditions considered by Pirmohammad and Kiani [31], as discussed above.

Figure 3.20 shows the fracture resistance of HMA mixtures modified by SBS and CR at different climate and loading conditions. Similar to the results of plain HMA mixture investigated by Pirmohammad and Kiani [31], fracture resistance of the modified HMA mixtures conditioned at the VT was also less than the CT under any mode of loading. In addition, the minimum value of fracture resistance was achieved under mixed mode of $M^e = 0.5$.

3.8 Effect of Nanomaterials

Due to the improved performance of asphalt concretes, the use of nanomaterials, such as Nano silicon dioxide and Nano titanium [52], Nano zinc oxide [53], Nano clay [54], carbon nanofibers [55], etc, has been paid attention by researchers in recent years. For example, Shafabakhsh and Ani [56] showed that Nano TiO_2 and Nano SiO_2 improve the rutting and fatigue resistances of asphalt concretes. Ashish et al. [57] added Nano clays to asphalt concretes and showed that all the moisture, rutting, and fatigue resistances are improved. In another study, Zhang et al. [58] added both nanoparticles and polymers (*i.e.*, Nano ZnO, Nano TiO_2, Nano $CaCO_3$, SBS, and SBR) to asphalt concrete. The results exhibited that they improve both the high- and low-temperature properties of asphalt mixtures simultaneously. According to Amin et al. [59], the rutting and fatigue resistances of unaged and aged binders increase as the carbon nanotubes (CNTs) content in the binder increases. Arabani and Faramarzi [60] showed that the addition of CNTs to binder improves the fatigue behavior significantly. Based on an investigation performed by Ziari et al. [61], the complex modulus, fatigue, and rutting resistances of asphalt concrete increase by modifying binders with the CNTs. In another study, Xiao et al. [62] used the CNTs as binder

Fig. 3.20 Fracture
resistance of HMA mixtures
modified by **a** SBS and **b** CR
at different climate and
loading conditions [51]

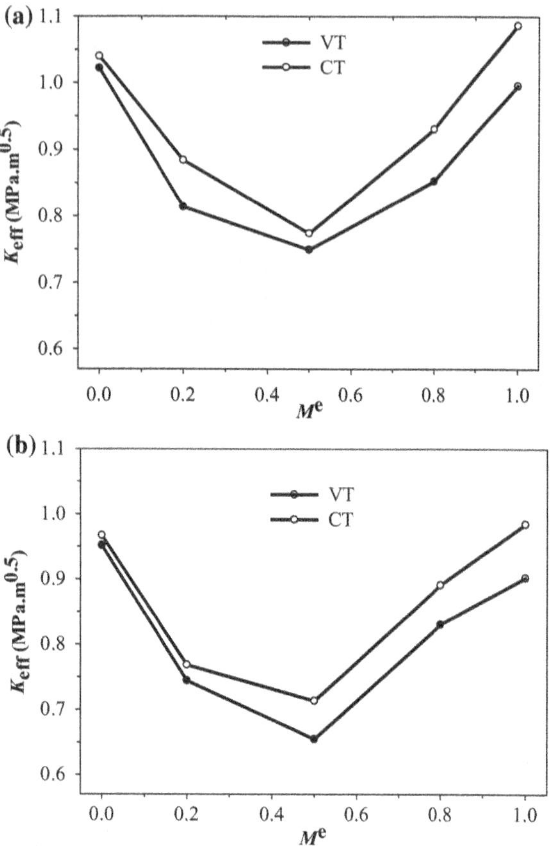

modifier and showed that the CNTs improve the fatigue and rutting resistances as
well as the linear viscoelastic response of binder. Kordi and Shafabakhsh [63] also
showed that the Nano Fe_2O_3 can improve the mechanical properties, fatigue, and
rutting resistances of asphalt concrete.

A few investigations can be found in the literature on fracture behavior of asphalt
concretes modified by nanomaterials. For example, Pirmohammad et al. [64] investigated mixed mode I/II fracture behavior of HMA mixtures by using the CNTs and
Nano Fe_2O_3 as binder modifiers.

Binder with penetration grade of 60/70 and aggregates with gradation No. 5,
presented in Table 3.2, were used for preparing the HMA mixtures. The NMAS
was selected 12.5 mm in this study because the aggregates used in the mixture
must be smaller than the specimen dimensions to correctly characterize the HMA
properties, as suggested by other researchers (see *e.g.*, [65, 66]). The HMA mixtures
were prepared by modifying the binders with four different percentages (*i.e.*, 0.1,
0.4, 0.8, and 1.2% by weight of the base binder) of the Nano Fe_2O_3 and CNTs
particles. In order to compare the fracture properties of the HMA mixtures modified

Table 3.4 Finite element results [64]

Mode of loading	M^e	$S_1, S_2,$ and L (mm)	Y_{I}	Y_{II}
Pure mode I	1	33, 33 and 0	4.29	0.00
Mixed mode I/II	0.66	33, 13 and 1.5	1.63	0.95
Mixed mode I/II	0.32	33, 13 and 6.5	0.89	1.62
Pure mode II	0	33, 13 and 11	0.00	2.26

by nanoparticles with those of the plain HMA mixtures, another mixture, called normal HMA mixture, was also prepared in this study, without modifying the base binder. It is also pointed out that all the HMA mixtures contained the same air void content of 4%.

For preparation of the modified HMA mixtures, the Nano Fe_2O_3 and CNTs particles were firstly mixed with the base binder using a shear mixer operating at 1800 rpm for 1 h to achieve a suitable homogeneity. During the mixing process, the temperature was set at 120 °C using a hot plate device. The modified binders were then blended with the aggregates, and cylindrical samples with 50 mm in radius and 60 mm in height were finally compacted in the laboratory. The cylindrical samples were cut in several steps to produce the SCB specimens containing an edge crack of length a = 17 mm at suitable positions presented in Table 3.4. The results of finite element analyses including the positions of the supports (*i.e.*, S_1 and S_2) and crack (*i.e.*, L) together with the values of geometry factors (*i.e.*, Y_{I} and Y_{II}) for different modes of loading are given in Table 3.4.

In this research [55], fracture tests were conducted under pure mode I, pure mode II, and two different mixed modes of $M^e = 0.66$ and $M^e = 0.32$. In order to perform the fracture tests, the SCB specimens were initially put into a freezer fixed at -15 °C for 6 h to ensure all parts of the specimens have the same temperature. They were then placed upon the bottom supports of the UTM adjusted at the proper locations (*i.e.*, S_1 and S_2) given in Table 3.4, and the top fixture was finally moved downward at a displacement rate of 3 mm/min to load them. Fracture load P_{cr} was obtained from the load-*LLD* curve recorded from the fracture tests. It is notable that the fracture test set-up shown in Fig. 3.6 was used in this study. As mentioned in Sect. 3.2.3, the fracture strength K_{eff} of HMA mixtures can be calculated by putting the values of P_{cr}, geometry factors (*i.e.*, Y_{I} and Y_{II} given in Table 3.4), and dimensions of SCB into Eqs. 3.4–3.6.

Figure 3.21 exhibits fracture strength of the Nano Fe_2O_3 modified HMA mixtures. For all the modes of loading, by increasing the Nano Fe_2O_3 content in the mixture, the fracture strength firstly increased to reach its maximum value at 0.8%, and then decreased, signifying that there was an optimum value for addition of the Nano Fe_2O_3 to HMA concrete. Therefore, 0.8% was the suitable value of the Nano Fe_2O_3 content for adding to asphalt mixtures. Kordi and Shafabakhsh [63] also observed similar results. Based on their study, the addition of 0.9% Nano Fe_2O_3 to the stone mastic asphalt mixtures showed the best results for the mechanical properties of asphalt mixtures while addition of higher amounts of the Nano Fe_2O_3 aggravated the

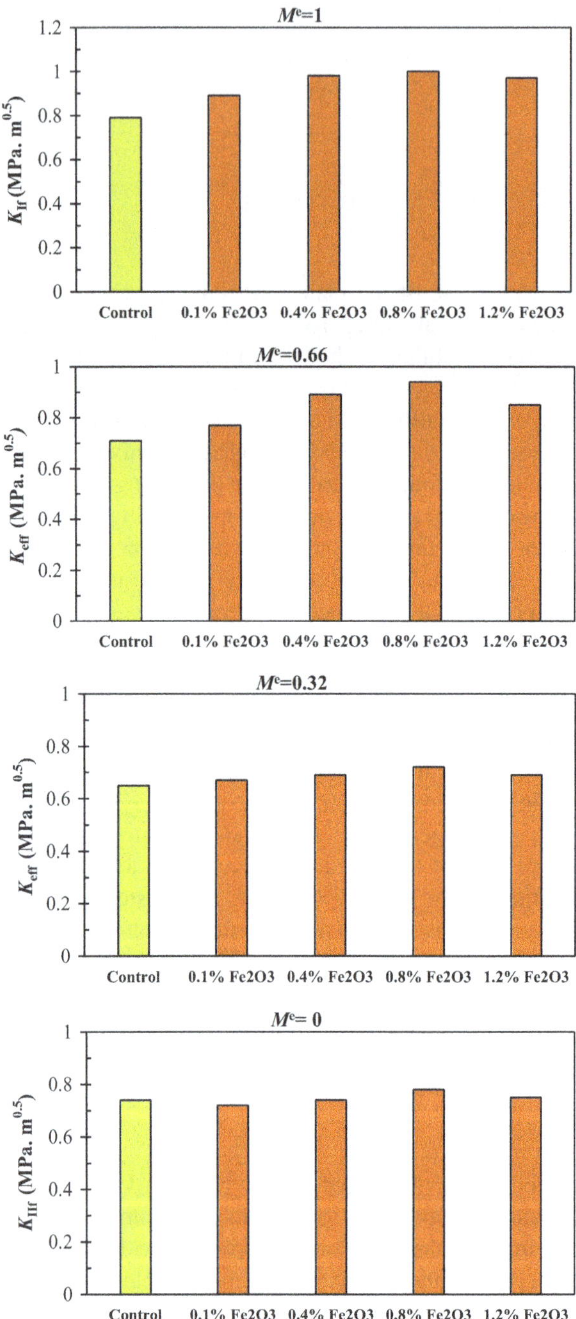

Fig. 3.21 Fracture strength of HMA mixtures modified by different percentages of Nano Fe_2O_3 under different modes of loading [64]

performance of mixture. Indeed, the Nano Fe_2O_3 contributed the binder particles to exhibit higher adhesion features, and therefore, the aggregates were strongly bonded together.

For the mixed mode of $M^e = 0.66$, the addition of 0.8% Fe_2O_3 to HMA mixture improved the fracture strength by 32%. Generally, an improvement of 13–27% for the pure mode I, 8–32% for the mixed mode of $M^e = 0.66$, and 3–11% for the mixed mode of $M^e = 0.32$ was observed depending on the amount of Nano Fe_2O_3 particles used in the mixture. Furthermore, the use of nanoparticles resulted in a negligible or negative effect on the fracture strength when the SCB specimens were loaded under pure mode II (see Fig. 3.21). Hence, for this case of loading, the best improvement in fracture strength was 5% for the HMA mixtures modified by 0.8% Nano Fe_2O_3.

The CNTs are expected to have a great deal of potential for improving the performance of HMA mixtures due to having outstanding mechanical properties [67, 68]. Furthermore, other interesting properties of CNTs such as high aspect ratio, specific strength, chemical resistance, electrical and thermal conductivity have made them to be used as a popular binder modifier [69, 70].

Figure 3.22 shows fracture strength of the HMA mixtures modified by CNTs under different modes of loading. According to the results, application of the CNTs improved the fracture strength of HMA mixtures. The CNT-modified binders are usually more viscous than the plain binders, tending to show an improved adhesive bonding for aggregating the particles. Moreover, better interlock between the CNT-modified binders and aggregates imparted strength to the mixture, leading to improve the fracture resistance [56].

According to the results, for all the modes of loading, the fracture strength increased as the amount of the CNTs in the mixture enhanced. In addition, improvement in fracture strength was more pronounced as the SCB specimens were loaded under dominant mode I (*i.e.,* pure mode I and mixed mode of $M^e = 0.66$). In other words, by increasing the amount of shear mode, the improvement effect of the CNTs on the fracture strength decreased gradually. As a conclusion, the highest improvement in the fracture strength (*i.e.,* 46%) was achieved when the HMA mixtures were modified by 1.2% CNTs and loaded under pure mode I. Generally, an improvement of 20–46% for the pure mode I, 14–38% for the mixed mode of $M^e = 0.66$, 3–20% for the mixed mode of $M^e = 0.32$, and 0–11% for the pure mode II was observed depending on the amount of CNTs particles used in the HMA mixture. The results also revealed that although the Nano Fe_2O_3 improved fracture strength of HMA mixtures significantly, however, the fracture strength of HMA mixtures was more improved by the CNTs than the Nano Fe_2O_3.

In another study, Pirmohammad et al. [71] evaluated fracture strength of HMA mixtures modified by Nanoclay and Nano Al_2O_3 particles at $-15\ ^\circ$C. To prepare the HMA mixtures, binder with penetration grade of 60/70 and aggregates with gradation No. 5, given in Table 3.2, were employed. The NMAS for this aggregate gradation was 12.5 mm. The HMA mixtures were prepared by modifying the binders with five different percentages (*i.e.,* 2, 2.5, 3, 3.5, and 4% by weight of the base binder) of the Nanoclay and four different percentages (*i.e.,* 0.1, 0.4, 0.8, and 1.2% by weight of the base binder) of the Nano Al_2O_3 particles.

Fig. 3.22 Fracture strength of HMA mixtures modified by different percentages of CNTs under different modes of loading [64]

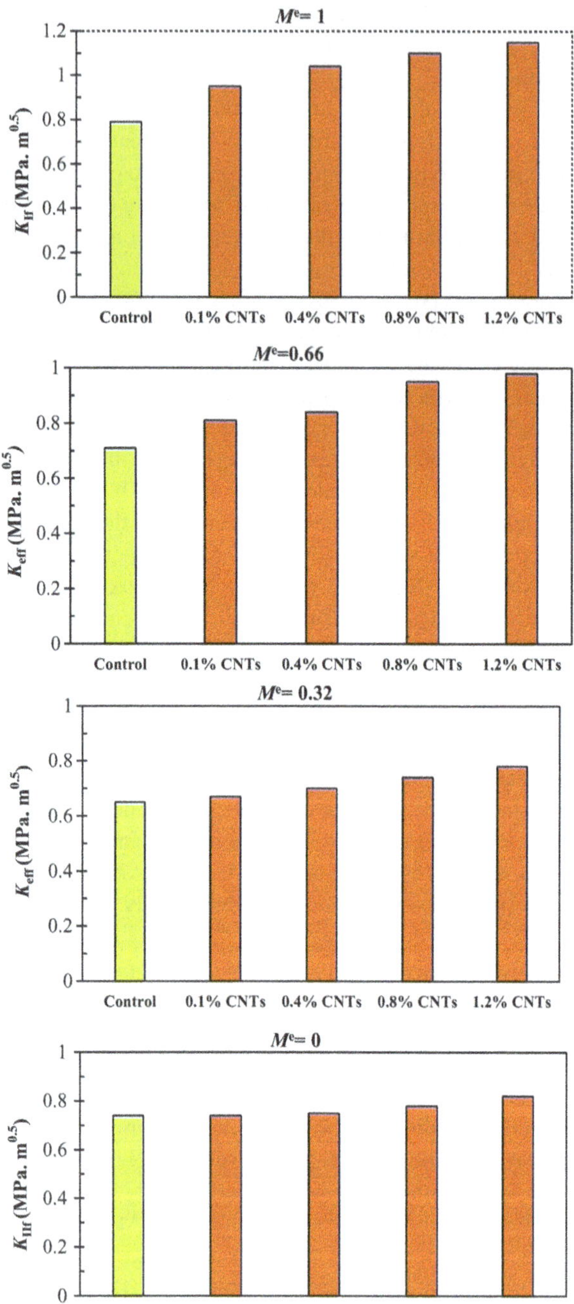

In order to compare the fracture properties of the HMA mixtures modified by nanoparticles with those of the plain HMA mixtures, another mixture, called normal HMA mixture, was also prepared. Moreover, all the HMA mixtures contained the same air void content of 4%.

In order to prepare modified HMA mixtures, the Nanoclay and Nano Al_2O_3 particles were initially mixed with the base binder using a shear mixer operating at 1800 rpm for 1 h to achieve a suitable homogeneity. During the mixing process, the temperature was set at 120 °C using a hot plate device. The modified binders were then blended with the aggregates, and cylindrical samples with 50 mm in radius and 60 mm in height were finally compacted in the laboratory. The cylindrical samples were cut in several steps to produce the SCB specimens containing an edge crack of length $a = 17$ mm at suitable positions presented in Table 3.4. The results of finite element analyses including the positions of the supports (*i.e.*, S_1 and S_2) and crack (*i.e.*, L) together with the values of geometry factors (*i.e.*, Y_I and Y_{II}) for different modes of loading are given in Table 3.4.

It is worth noting that a similar procedure used by Pirmohammad et al. [64] for conducting the experiments was employed in this study, as explained earlier.

Figure 3.23 exhibits the fracture strength of Nanocaly modified HMA mixtures subjected to different modes of loading. The results indicated that the Nanoclay particles improved the fracture strength of HMA mixtures. The similar results have been observed by You et al. [72], in which the addition of Nanocaly particles to the binder increased the toughness of binder. It is mentioned that the toughness of binder is measured by calculating the area under the stress–strain curve obtained from the direct tensile tests. Furthermore, the Nanoclay particles affected the rheological behavior and internal structure of the base binder. The elasticity, stiffness, and aging resistance of the Nanoclay-modified binder were much higher than those of the plain one [54]. Hence, improvement in the fracture strength of the Nanoclay-modified HMA mixtures corresponds to the stiffening effect of the Nanoclay particles as they form bond chains within the binder.

The results showed that the addition of Nanoclay increased fracture strength of the HMA mixtures under all the modes of loading. But its improvement effect was more pronounced as the SCB specimens were subjected to the pure mode II and the mixed mode of $M^e = 0.32$. In other words, the Nanoclay particles more improved the fracture strength of HMA mixtures when the amount of shear mode at the crack front enhanced. On average, the use of Nanoclay particles with the percentages of 2, 2.5, 3, 3.5, and 4 increased the fracture strength by 2%, 8%, 14%, 19%, and 31%, respectively. Therefore, the highest improvement in fracture strength was achieved by addition of 4% Nanoclay to the HMA mixture. Particularly, the use of 4% Nanoclay particles in the HMA mixture increased the fracture strength by 38% for the mixed mode of $M^e = 0.32$ and 35% for the pure mode II. It was also found that low contents of the Nanclay particles (less than 3%) approximately improved the fracture strength identically for all the modes of loading.

Figure 3.24 shows the fracture strength of Nano AL_2O_3 modified HMA mixtures under different modes of loading. According to this figure, the fracture strength gradually increased as the percentage of Nano AL_2O_3 particles enhanced. Based on

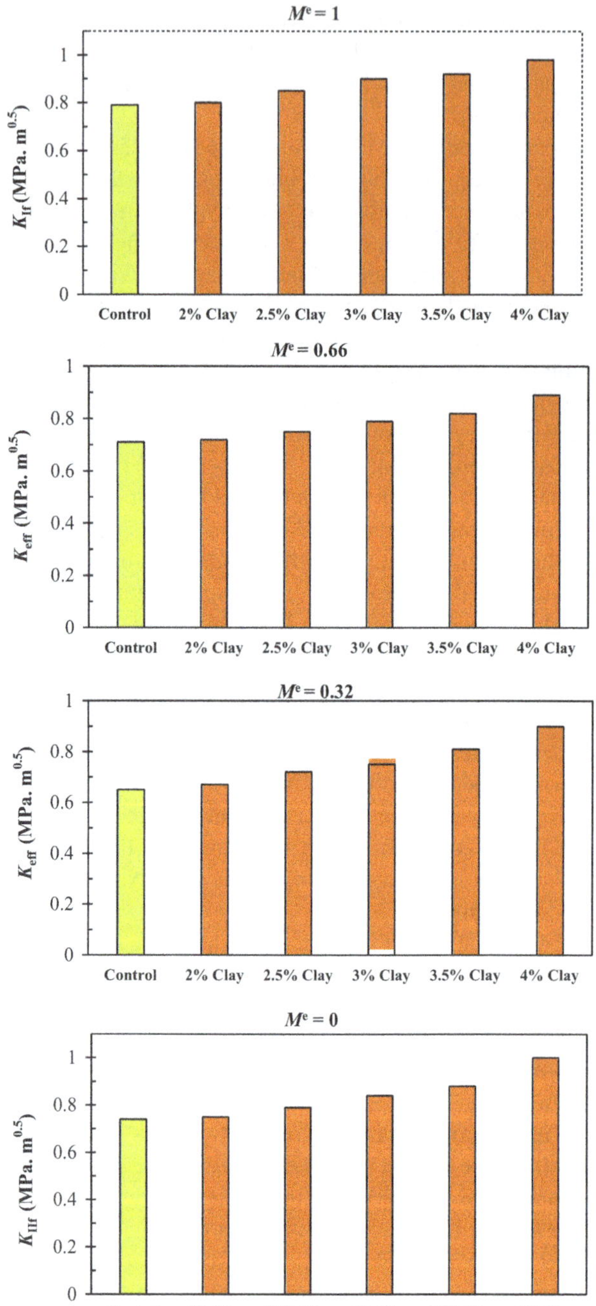

Fig. 3.23 Fracture strength of HMA mixtures modified by different percentages of Nanoclay under different modes of loading [71]

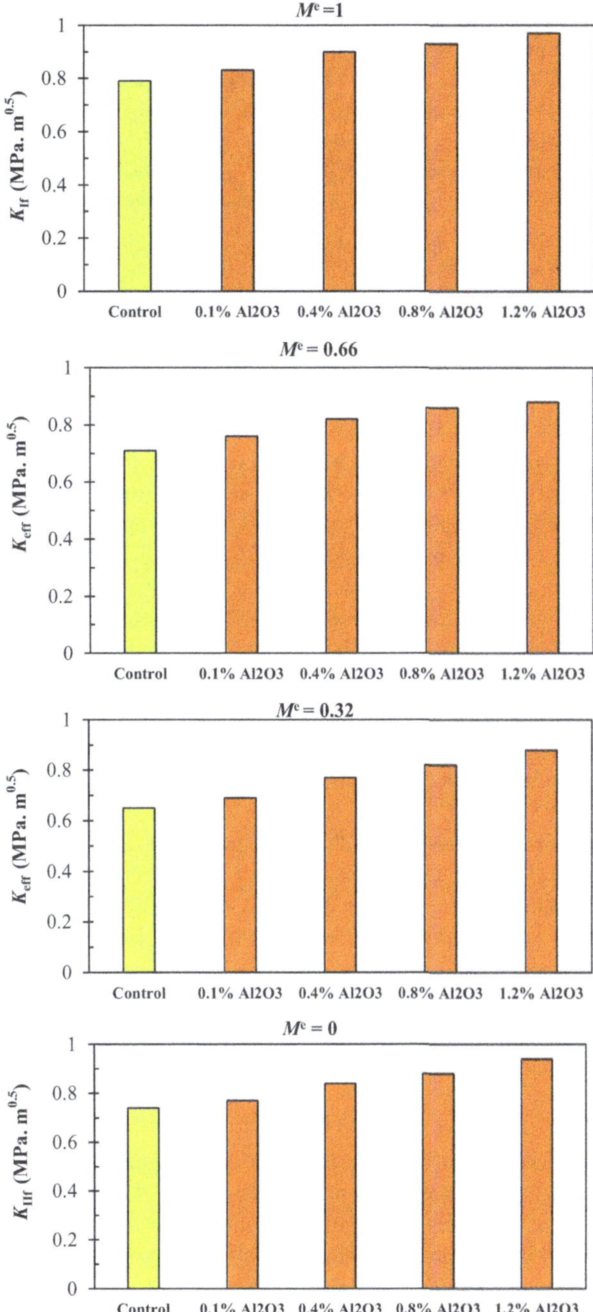

Fig. 3.24 Fracture strength of HMA mixtures modified by different percentages of Nano Al_2O_3 under different modes of loading [71]

an investigation performed by Al-Mansob et al. [73], the adhesion force of binder increased as the amount of Nano AL_2O_3 in the HMA mixture increased. This indicates that good adhesion was formed between the aggregates in the HMA mixtures containing Nano AL_2O_3. Furthermore, the Nano AL_2O_3 increased the viscosity, stiffness, and elastic behavior of the binder.

Similar to the results observed for the Nanaoclay-modified HMA mixtures, improvement in the fracture strength was more pronounced as the portion of shear mode was dominant at the crack front (*i.e.*, $M^e = 0.32$ and pure mode II). On average, the use of Nano AL_2O_3 with the percentages of 0.1, 0.4, 0.8, and 1.2 increased the fracture strength by 6%, 15%, 21%, and 27%, respectively. Particularly, the use of 1.2% Nano AL_2O_3 particles in the HMA mixture increased the fracture strength by 35% for the mixed mode of $M^e = 0.32$. In addition, similar to the results observed for the Nanoclay particles, low contents of the Nano AL_2O_3 particles (less than 0.4%) approximately improved the fracture strength identically for all the modes of loading.

Ameri et al. [74] investigated the effect of CNTs on the fracture properties of asphalt concrete using the SCB tests. Binder with penetration grade of 60/70 and aggregate gradation No. 4 (presented in Table 3.2) were employed in this research. The CNT particles with various dosages of 0, 0.2, 0.5, 0.8, 1.2, and 1.5% by weight of the binder were added to the binder. The addition of the CNT particles to the binder was conducted using an ultrasonic mixer for 15 min to suitably disperse the nanoparticles within the binder for avoiding defects caused by agglomeration of the CNT particles.

They used the *J*-integral concept to investigate the fracture properties of HMA mixtures. The *J*-integral is a path-independent method characterizing the energy release rate for nonlinear fracture behaviors. Details related to the *J*-integral method have been given in Chap. 5.

For each type of asphalt concrete, the SCB specimens with three different crack lengths of 25.4, 31.8, and 38.1 mm in the middle of the specimen were provided. The specimens had a diameter of 150 mm and thickness of 30 mm. It is reminded that the preparation of specimens with at least two different crack lengths is essential in the *J*-integral method.

Fracture tests were conducted on the SCB specimens using a three-point bend fixture with a displacement rate of 0.5 mm/min at the temperature of 20 °C. Three replicates for each crack length were tested for reducing the discrepancies in the results. They found the value of J_c as 0.1667, 0.1867, 0.3867, 0.5367, 1.14, and 2.7833 for the CNTs dosages of 0, 0.2, 0.5, 0.8, 1.2, and 1.5%, respectively. According to the results, fracture strength of the asphalt concrete increased as a dosage of CNTs in the asphalt concrete enhanced. Furthermore, the effect of carbon nanotube became more pronounced as its percent content within the binder increased.

Aliha et al. [75] investigated the effect of Nanosilica and Nanoclay on the mode I fracture toughness of asphalt concrete using a disk-shaped specimen shown in Fig. 3.25.

A base binder with penetration grade of 60/70 and limestone aggregates with gradation No. 3 (presented in Table 3.2) were utilized to prepare the HMA mixtures using a superpave gyratory compactor. All the mixtures had an air void content of

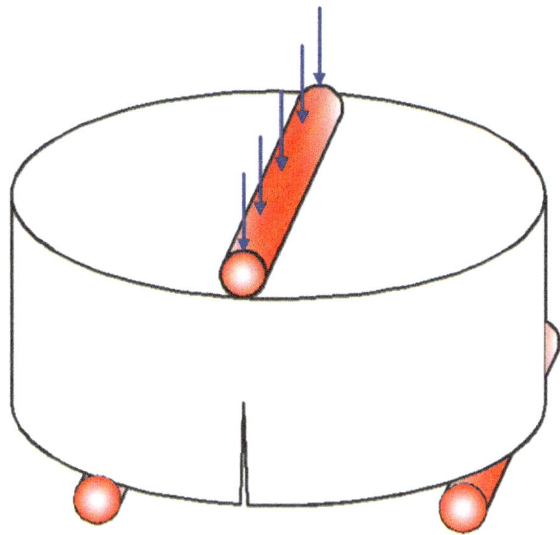

Fig. 3.25 Geometry of disk-shaped specimen [75]

4.5%. The disk-shaped specimens had a diameter of 150 mm, height of 30 mm, and crack length of 12 mm. It is also pointed out that in order to prepare the Nanoclay-modified binder, the base binder was mixed with 4% (by weight) Nanoclay particles at 150 °C for 20 min using an ultrasonic device with the power of 65 kW. As such, the Nanosilica-modified binder was prepared by mixing 4% Nanosilica with the base binder at 175 °C for 2 h using a high shear mixer working at 4000 rpm.

Fracture tests were conducted on the specimens with a span (*i.e.*, distance between the bottom supports) of 135 mm and displacement rate of 3 mm/min using a three-point bend fixture at −15 °C. Figure 3.26 shows the fracture toughness of the modified asphalt concretes, in which the nanomaterials had positive effect on enhancing the fracture resistance of asphalt concrete. In addition, the Nanoclay-modified asphalt

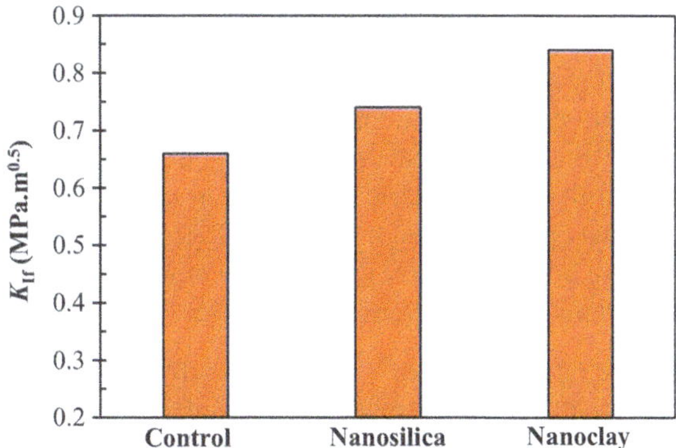

Fig. 3.26 Mode I fracture toughness of HMA mixtures [75]

concrete demonstrated the highest fracture toughness (*i.e.,* 30% greater than the control HMA) compared to the control and Nanosilica modified HMA concretes.

3.9 Effect of Fibers

Fibers are frequently used as a reinforcing material in asphalt pavements. Fibers are generally classified into two groups: (i) synthetic (*i.e.,* human-made) fibers and (ii) natural (*i.e.,* plant and animal) fibers [76]. Researchers employed different types of fiber such as polypropylene [77], polyester [78], polyethylene terephthalate [79], steel [80], carbon [81], glass [82], etc. in the past years to improve performance of asphalt concrete. For example, Klinsky et al. [83] incorporated a mixture of aramid and polypropylene fibers into asphalt concrete to better its performance. Apostolidis et al. [84] utilized two synthetic fibers including aramid and polyolefin to improve mechanical characteristics of asphalt mortar.

On the other hand, researchers have also used natural fibers as an additive in asphalt concretes. Natural fibers are environmentally friendly because of possessing no or very low content of CO_2 in the production processes. Meanwhile, the natural fibers like jute, kenaf, sisal, coconut, etc. absorb the CO_2 from the atmosphere. According to Sani et al. [85] and Mughal et al. [86], by addition of kenaf and coir fibers to asphalt mixture, its Marshall stability was improved. In addition, application of kenaf fibers mitigated both the rutting and moisture damages [87].

There are few investigations in the literature studying the effect of fibers on the fracture behavior of HMA mixtures, which are discussed herein.

Pirmohammad et al. [88–90] used four different fibers (*i.e.,* kenaf, goat wool, basalt, and carbon) as additives to evaluate fracture resistance of HMA mixture. Binder with penetration grade of 60/70 (as a base binder) and aggregates with gradation No. 5, presented in Table 3.2, were used for preparation of HMA concretes. In this study, three different percentages (*i.e.,* 0.1, 0.2, and 0.3% by weight of asphalt mixture) as well as three different lengths (*i.e.,* 4, 8 and 12 mm) of fibers were used to prepare the reinforced HMA concretes. In order to compare the fracture strength of the reinforced HMA concretes with that of the plain HMA concrete, another mixture (*i.e.,* control HMA concrete) was also prepared. It is pointed out that all the HMA mixtures contained the same aggregate gradation of No. 5 and air void of 4%.

Fibers are recommended to be dried [91], so the kenaf, goat wool, basalt, and carbon fibers were dried by heating in an oven for one hour at 160 °C. Table 3.5 presents the properties of fibers employed in this study.

Researchers use three methods of wet, dry, and combination of them to incorporate fibers into asphalt concretes:

Wet method: Fibers and binder are initially blended using a high shear mixer, and the blend is then mixed with aggregates (see, *e.g.,* [81, 91]).
Dry method: Fibers and aggregates are initially blended, and the blend is then mixed with binder. This method is preferable to the wet method because it is simple and

Table 3.5 Properties of the fibers used in this study [88–90]

Fiber type	Elastic modulus (GPa)	Tensile strength (MPa)	Diameter (μm)	Density (g/cm^3)
Kenaf	38	701	70–75	0.94
Goat wool	19	1580	60–75	1.28
Basalt	85	2800	10–13	2.67
Carbon	238	4300	5–6	1.76

does not need high shear mixer. Furthermore, agglomeration of fibers, which generates weak zones within mixture, in the dry method is much less than the wet one [79, 92, 93].

Combination of wet and dry methods: Aggregates and binders are previously blended before adding fibers to the mixture (see *e.g.*, [94]).

The third method (*i.e.*, combination of wet and dry methods) was used in this study because it showed the least agglomeration of fibers compared to the wet and dry methods. To further clarify this method, it is noticed that aggregates and binder were blended for 3 min at 145 °C, and the fibers were then dispersed into the mixture gradually. The mixture was finally blended for 5 min to obtain a uniform dispersion of fibers within the mixture. It is worth mentioning that long fibers and high contents of fibers may lead to agglomeration problems.

Cylindrical samples with 50 mm in radius and 60 mm in height were compacted by the Marshall compactor machine with 75 blows on the top and bottom ends. The cylindrical samples were cut in several steps to produce the SCB specimens containing an edge crack of length $a = 17$ mm at suitable positions presented in Table 3.4. The results of finite element analyses including the positions of the supports (*i.e.*, S_1 and S_2) and crack (*i.e.*, L) together with the values of geometry factors (*i.e.*, Y_I and Y_II) for different modes of loading are given in Table 3.4.

Fracture experiments were carried out under pure mode I ($M^\mathrm{e} = 1$), pure mode II ($M^\mathrm{e} = 0$), and two different mixed modes of $M^\mathrm{e} = 0.66$ and $M^\mathrm{e} = 0.32$ at −15 °C. Furthermore, the SCB tests were performed at a displacement rate of 3 mm/min. Fracture load P_cr was then obtained from the load-*LLD* curve recorded using the test set-up shown in Fig. 3.6. As mentioned in Sect. 3.2, the fracture strength K_eff of HMA mixtures can be calculated by putting the values of P_cr, geometry factors (*i.e.*, Y_I and Y_II given in Table 3.4), and dimensions of SCB into Eqs. 3.4–3.6.

Figure 3.27 shows the fracture resistance of the kenaf reinforced HMA concretes subjected to different modes of loading. Based on the results, the fracture resistance of the kenaf reinforced HMA concretes was greater than that of the control HMA concrete, indicating that the kenaf fibers increased the fracture resistance. Indeed, the fibers provided additional tensile strength and therefore improved the fracture resistance of asphalt mixtures. Furthermore, the photographs taken using scanning electron micrographs (SEM) indicated that the surface of kenaf fibers has longitudinal ridges which can interlock with the binder matrix and therefore increase the fracture resistance of asphalt concrete.

Fig. 3.27 Fracture strength of the kenaf reinforced HMA concretes under different modes of loading [88]

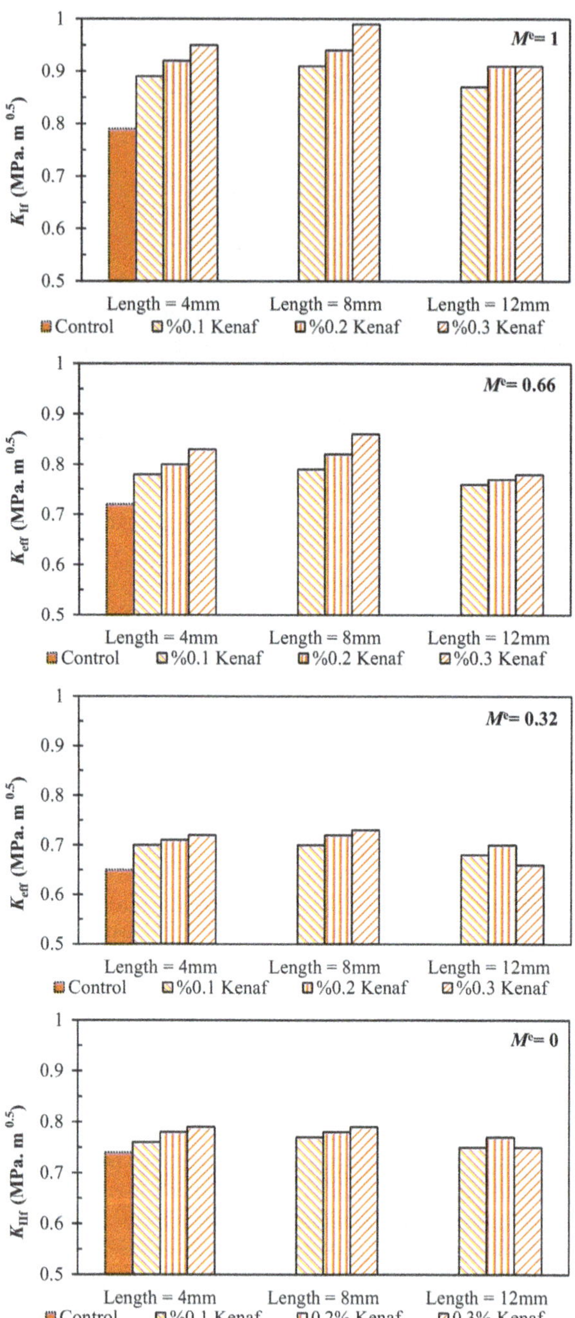

Fracture resistance of HMA concretes containing the kenaf fibers with the lengths of 4 and 8 mm increased as the fiber dosage enhanced. While a different behavior was observed for the kenaf length of 12 mm, *i.e.*, the fracture resistance firstly increased and then nearly remained constant or decreased, depending on the mode of loading.

Figure 3.28 displays the amount of improvement in fracture resistance for the kenaf reinforced HMA concretes. The improvement of fracture resistance initially increased and then decreased as the kenaf length enhanced. Hence, 8 mm kenaf fiber showed the highest positive effect on the fracture resistance for any dosage of kenaf fibers available in the mixture. Furthermore, the positive effect of the kenaf fiber decreased as the proportion of mode II relative to mode I (at the crack front of the SCB specimen) increased. Accordingly, the highest amount of improvement in fracture resistance was observed for the case of pure mode I, while a slight improvement in

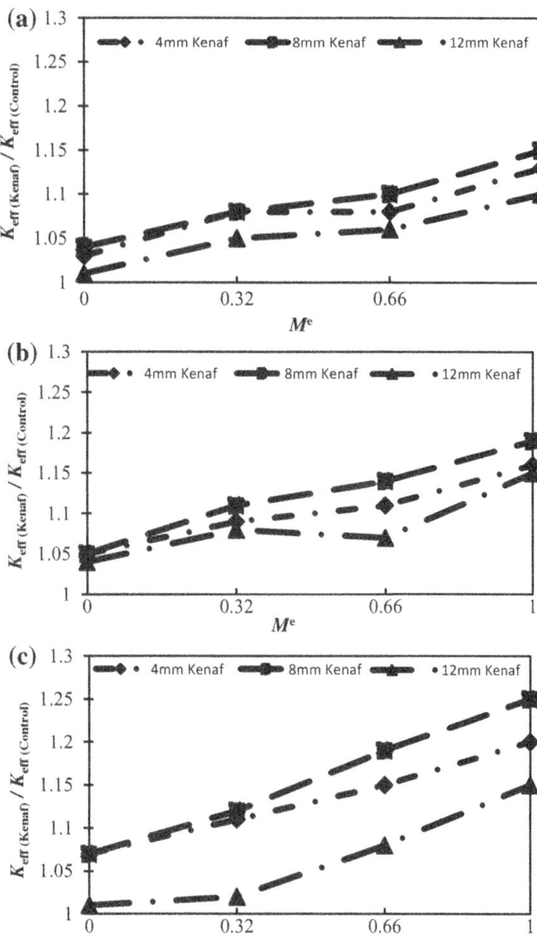

Fig. 3.28 Ratio of K_{eff} for the kenaf reinforced HMA concrete to that for the control HMA concrete with different fiber dosages of **a** 0.1%, **b** 0.2%, **c** 0.3% [88]

fracture resistance was observed as the HMA concretes were loaded under pure mode II. Similar results have been reported by Aliha et al. [76]. Their results showed that the jute and FORTA (polyolefin–aramid) fibers have increased the mode I fracture resistance of warm mix asphalt concretes more than the mode II one. In other words, for the loadings that the proportion of mode II was considerable (*i.e.,* for smaller values of M^e), none of the fibers affected the fracture resistance remarkably.

The results given in Fig. 3.28 revealed that the 8 mm kenaf fiber with a dosage of 0.3% demonstrated the highest positive effect on the fracture resistance of HMA concrete compared to other mixtures reinforced by kenaf fibers. This mixture showed 25, 19, 12, and 7% higher fracture resistance than the control HMA concrete under pure mode I, mixed mode of $M^e = 0.66$, mixed mode of $M^e = 0.32$ and pure mode II, respectively.

Figure 3.29 exhibits the effect of goat wool fibers on the fracture resistance of HMA concretes under different modes of loading. According to the results, the goat wool fiber improved the fracture resistance of HMA concrete considerably. For the goat wool fibers with the lengths of 4 mm and 8 mm, the fracture resistance improved as the fiber dosage enhanced. For the 12 mm goat wool fibers, the fracture resistance firstly increased and then decreased.

According to Fig. 3.30, the amount of improvement in the fracture resistance of the goat wool reinforced HMA concretes reduced as the fiber length enhanced. Hence, 4 mm goat wool fiber demonstrated more desirable results compared to longer fibers. In addition, the positive effect of the goat wool fibers on the fracture resistance of mixtures was approximately identical for all the modes of loading. This may be imputed to the fact that the goat wool fibers were soft and were not broken during the pulling-out process. Indeed, the goat wool fibers bridged the crack (see Fig. 3.31a), leading to improve the fracture resistance as the specimens were subjected to mode II loading. On the contrary, other fibers used in this study were broken during the crack opening process due to their brittle behavior, as shown in Fig. 3.31b.

The results given in Fig. 3.30 exhibited that the 4 mm goat wool fiber with a dosage of 0.3% had the highest positive influence on the fracture resistance of HMA concrete compared to other lengths and dosages of the goat wool fibers. This mixture had about 21% greater fracture resistance than the control HMA concrete.

Figure 3.32 exhibited the fracture resistance of basalt reinforced HMA concretes subjected to different modes of loading. According to this figure, application of basalt fibers increased the fracture toughness of HMA concretes. Moreover, for all the modes of loading, the fracture resistance increased as the fiber dosage in the mixture increased. Therefore, the use of 0.3% basalt fiber provided the highest fracture resistance. On the other hand, application of longer basalt fibers in the mixture resulted in reduction of the fracture resistance. Hence, 4 mm basalt fibers provided higher fracture resistance than 8 and 12 mm ones. This trend may be attributed to agglomeration of long fibers, resulting in weaker planes in the mixture. Similar results have been reported in other researches (see *e.g.,* [95, 96]), in which the modulus of rupture decreased as the length of fiber enhanced. Particularly, according to Qin et al. [97], the crack resistance and strength of HMA concretes reinforced by 6 mm basalt fibers were higher than those reinforced by 9 and 15 mm basalt fibers.

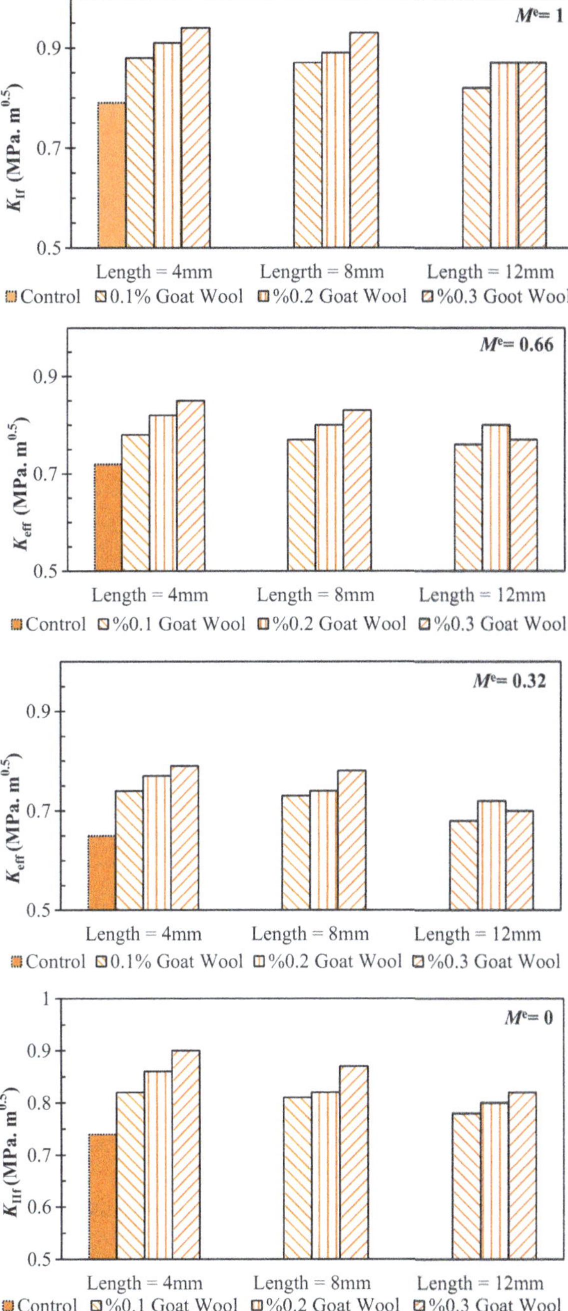

Fig. 3.29 Fracture strength of the goat wool reinforced HMA concretes under different modes of loading [88]

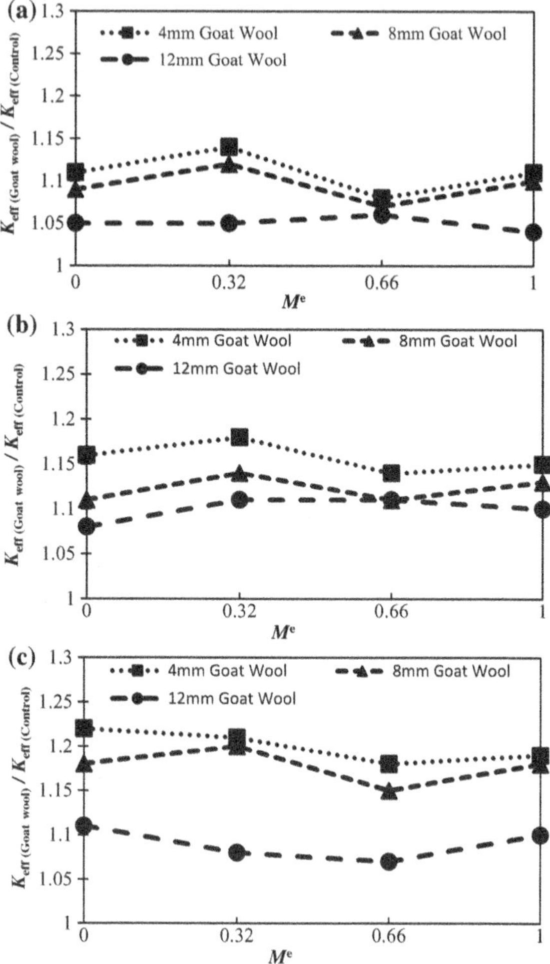

Fig. 3.30 Ratio of K_{eff} for the goat wool reinforced HMA concrete to that for the control HMA concrete with different fiber dosages of a 0.1%, b 0.2%, c 0.3% [88]

Based on Fig. 3.33, the positive effect of the basalt fiber reduced as the proportion of mode II at the crack front of the SCB specimen increased. Accordingly, the highest amount of improvement in fracture resistance was observed for the case of pure mode I, while a slight improvement in fracture resistance was achieved as the HMA concretes were loaded under pure mode II.

From the abovementioned explanations, it can be concluded that the 4 mm basalt fiber with a dosage of 0.3% demonstrated the best performance among the basalt fibers considered in this study. This mixture provided 28, 21, 15, and 16% greater fracture resistance than the control HMA concrete under pure mode I, mixed mode of $M^{e} = 0.66$, mixed mode of $M^{e} = 0.32$, and pure mode II, respectively.

Figure 3.34 shows the fracture resistance of HMA mixtures reinforced by carbon fibers with different lengths and dosages. Similar to the other fibers discussed above,

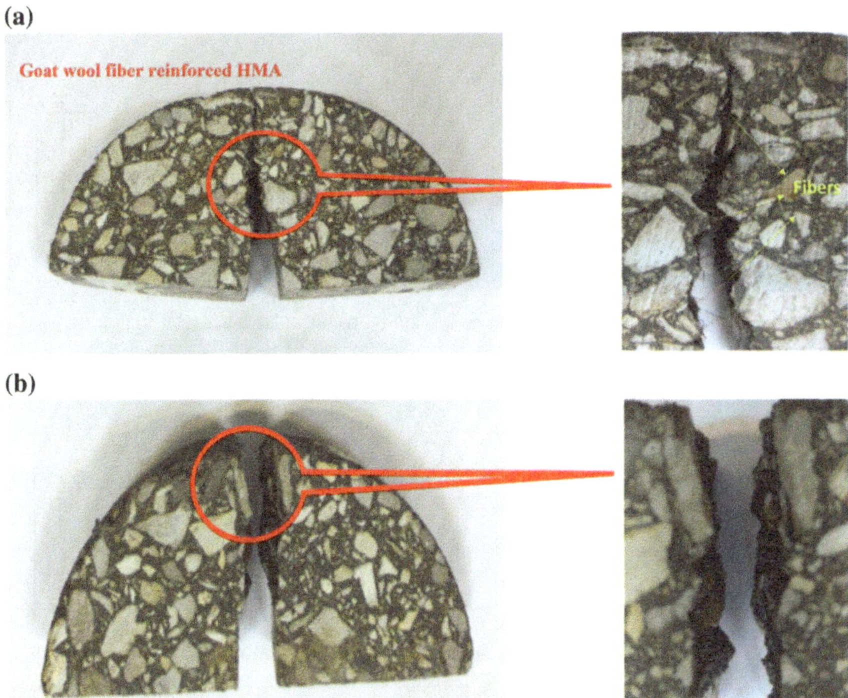

Fig. 3.31 **a** Crack bridging by goat wool fibers, **b** breakage of fibers during crack propagation in SCB specimens [88]

the carbon fibers also improved the fracture resistance of HMA concrete significantly. Since binder is a precursor of the carbon fiber, they are inherently compatible [98]. Moreover, the use of fibers imparted ductility and tensile strength to the mixture due to enhancing the interlock between aggregates. In the meanwhile, binder behaves as a brittle material at low temperatures, and therefore, the addition of fibers to asphalt concrete would act as a bridge [99, 100]. On the other hand, the SEM images taken from surface of the carbon fiber showed that its surfaces were not smooth and had many longitudinal ridges, interlocking with the binder matrix, and so increased the fracture resistance of HMA concrete.

Generally, for the 4 and 8 mm carbon-reinforced HMA concretes subjected to any mode of loading, the fracture resistance increased gradually as a dosage of fiber enhanced in the mixture. Meanwhile, the fracture resistance increased and then decreased (for the pure mode I and mixed mode of $M^{e} = 0.66$) and/or decreased (for the pure mode II and mixed mode of $M^{e} = 0.32$) when 12 mm carbon fibers were used in the mixture.

According to Fig. 3.35, when the amount of mode II at the crack front increased (*i.e.*, the value of M^{e} decreased), the positive effect of the carbon fiber reduced. As a result, the highest amount of improvement in fracture resistance was observed for

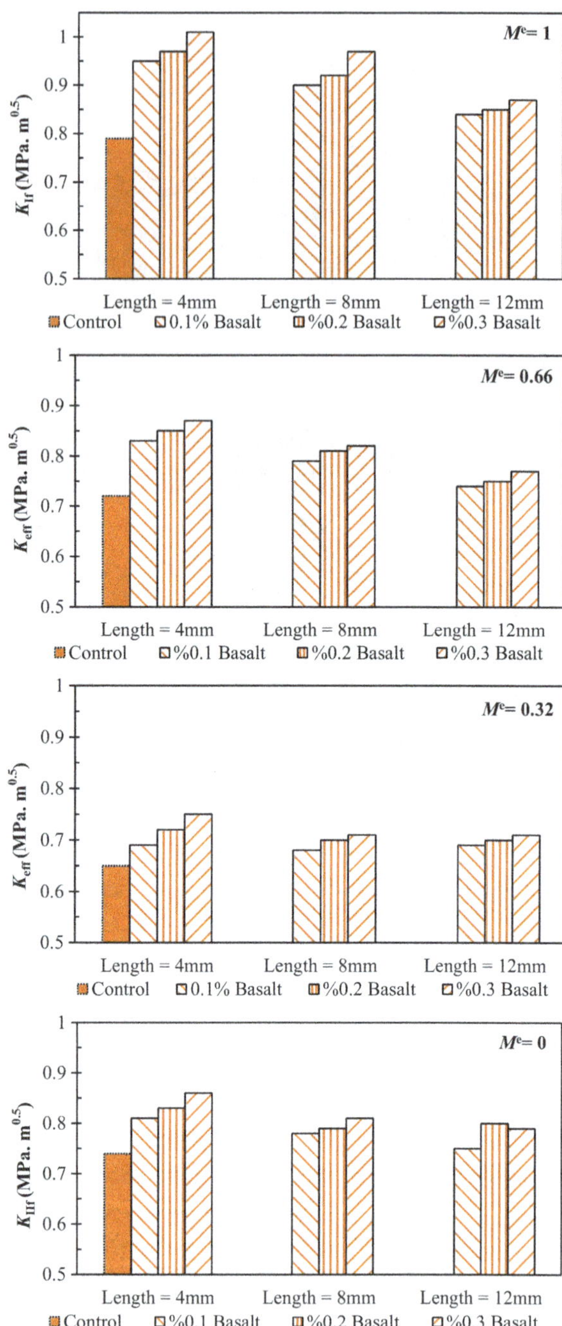

Fig. 3.32 Fracture strength of the basalt reinforced HMA concretes under different modes of loading [89]

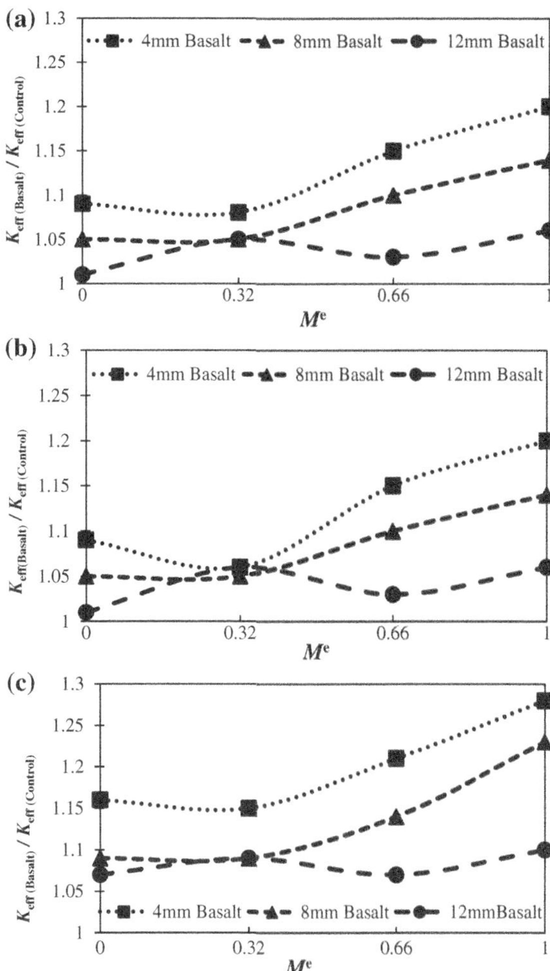

Fig. 3.33 Ratio of K_{eff} for the basalt reinforced HMA concrete to that for the control HMA concrete with different fiber dosages of **a** 0.1%, **b** 0.2%, **c** 0.3% [89]

the cases of mode I dominant loading (*i.e.*, pure mode I and mixed mode of $M^e =$ 0.66). This trend of the results may be attributed to the fact that the carbon fibers are brittle [101] and may break easily under shear loading, while they are strong under tensile (*i.e.*, mode I) loading.

Another point from Fig. 3.35 is that the longer carbon fiber with the lower dosage and the shorter carbon fiber with the higher dosage performed better than other compositions of the carbon fibers with different lengths and dosages. Apostolidis et al. [84] reported similar results when investigating the tensile strength and fatigue life of HMA concretes containing aramid and polyolefin fibers. They concluded that the longer fibers of low dosages generated equivalent performance to the HMA mixture containing shorter fibers of high dosages.

Fig. 3.34 Fracture strength of the carbon-reinforced HMA concretes under different modes of loading [90]

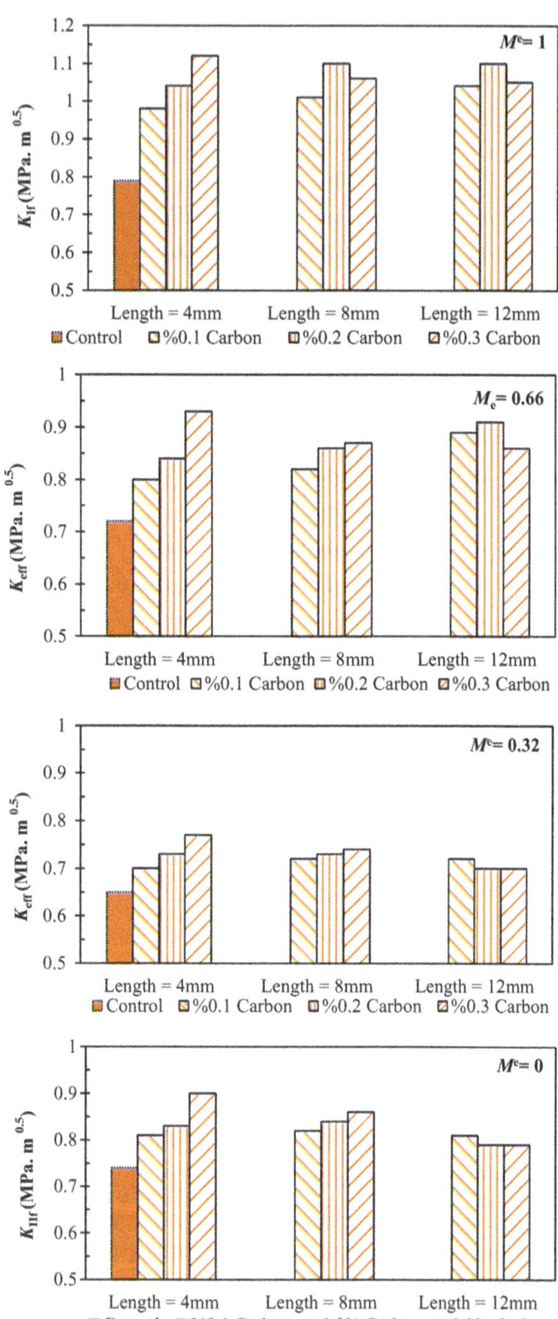

Fig. 3.35 Ratio of K_{eff} for the carbon-reinforced HMA concrete to that for the control HMA concrete with different fiber dosages of **a** 0.1%, **b** 0.2%, **c** 0.3% [90]

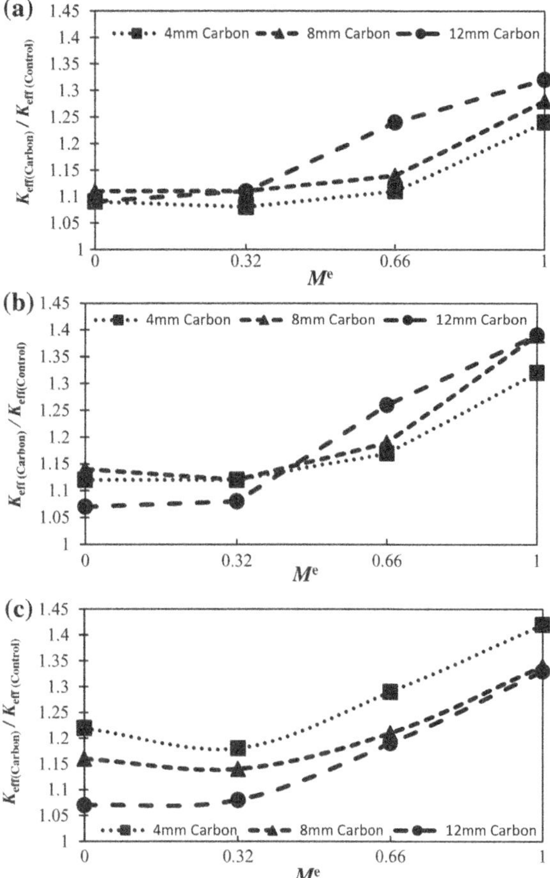

Based on the results given in Fig. 3.35, the 4 mm carbon fiber with a dosage of 0.3% demonstrated the greatest fracture resistance. This mixture had 42, 29, 18, and 22% higher fracture resistance than the control HMA concrete under pure mode I, mixed mode of $M^{\text{e}} = 0.66$, mixed mode of $M^{\text{e}} = 0.32$, and pure mode II, respectively. It is worth mentioning that the carbon fiber provided the highest positive effect on the fracture strength of HMA concrete compared to other fibers (*i.e.,* kenaf, goat wool, and basalt) investigated in this study.

The results also showed that the fracture resistance of all the mixtures investigated in this task initially decreased and then increased as the proportion of mode II enhanced (*i.e.,* the value of M^{e} decreased); as a result, all the mixtures had their minimum fracture resistance as they were subjected to the mixed mode of $M^{\text{e}} = 0.32$. Consequently, all the mixtures containing fibers were found to be more vulnerable to mixed mode I/II loading than pure modes of I and II.

3.10   Effect of Additives

Binder behaves as a brittle material when it is employed at low temperatures, and therefore, its brittle behavior results in appearance of premature cracking in the asphalt pavements. Additives have been shown to expand the operational temperature range of binder due to changing the softening temperature. Hence, modifying binder is one of the most common methods of improving the performance of asphalt mixtures [102].

Various binder additives such as styrene-butadiene-styrene (SBS), ethylene vinyl acetate (EVA), polyethylene (PE), crumb rubber (CR), etc. have been employed by researchers to improve performance of asphalt concretes (see *e.g.,* [30, 37, 103–105]). Some of the researches investigating the effect of additives on the fracture behavior of asphalt concretes are reviewed in this section.

Pirmohammad and Ayatollahi [30] investigated the effect of SBS on the fracture toughness of HMA mixture at different temperatures (*i.e.,* 0, −10, −20, and −35 °C). The experiments were conducted using the SCB specimen shown in Fig. 3.4a under pure mode I, pure mode II, and three mixed modes of $M^e = 0.8$, $M^e = 0.5$, and $M^e = 0.2$.

Figure 3.36 shows the fracture toughness of SBS modified HMA mixture, in which by reduction in the temperature, the fracture toughness initially increased and then decreased due to the DTC damage.

Figure 3.37 displays the ratio of K_{eff} for the SBS modified HMA mixture to that for the unmodified one at different temperatures and modes of loading. According to this figure, the K_{eff} ratio was greater than one for all the test temperatures and modes of loading, indicating that the SBS improved the fracture resistance of HMA mixture at any temperature and loading conditions. Interestingly, the effect of SBS was more pronounced as the HMA mixture was exposed to lower temperatures. Indeed, at low temperatures, the SBS imparted increased softness to the brittle binder and also provided improved adhesion between aggregates and binder [42, 43, 106]. As a result, it increased the fracture resistance of HMA mixture particularly at lower temperatures. In addition to the test temperature, the K_{eff} ratio was also dependent on the mode of loading. The SBS more influenced the fracture toughness of HMA

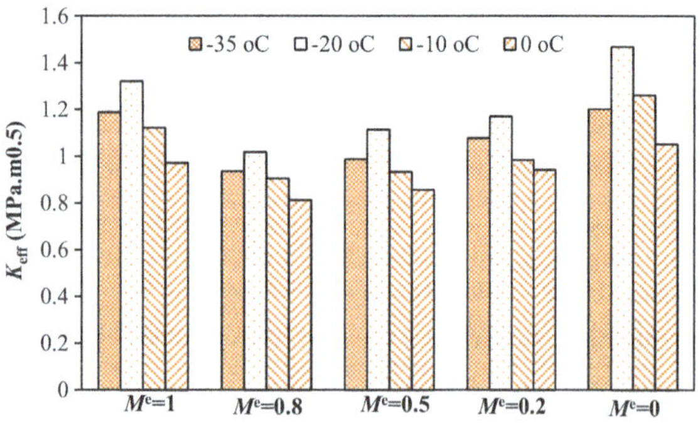

Fig. 3.36 Fracture toughness of HMA mixtures modified by SBS at various temperatures and modes of loading [30]

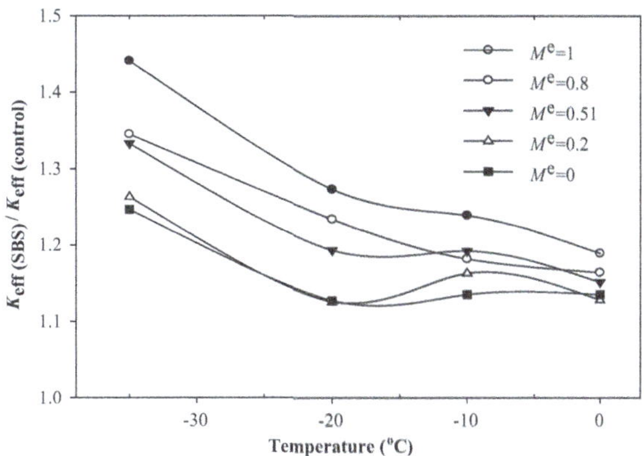

Fig. 3.37 Ratio of K_{eff} for the SBS modified asphalt concrete to that for the unmodified one at different temperatures and loading modes [30]

mixture under pure mode I than that under mixed mode and pure mode II, such that its positive effect on the mode I fracture toughness ranged from 19% (at 0 °C) to 44% (at −35 °C).

Pirmohammd et al. [13] measured mode I fracture energy of HMA concretes using the SCB specimen. Two types of HMA concretes were prepared in this study. One of them contained a binder with penetration grade of 60/70, and the second one contained 3.5% SBS by weight of a binder with penetration grade of 85/100. The mixtures also contained aggregate gradation No. 4, presented in Table 3.2, and air void of 4%.

The SCB specimens had a diameter of 150 mm, thickness of 32 mm, and crack length of 20 mm. The specimens were maintained at the temperature of −10 °C for 12 h. They were then loaded under pure mode I to record the load–*CMOD* curves. The value of *CMOD* was measured using a clip gage. Meanwhile, three SCB specimens were tested for each asphalt mixture, and the results were then averaged to obtain the fracture energy. The results are shown in Fig. 3.38, in which the fracture energy increased significantly by the use of SBS.

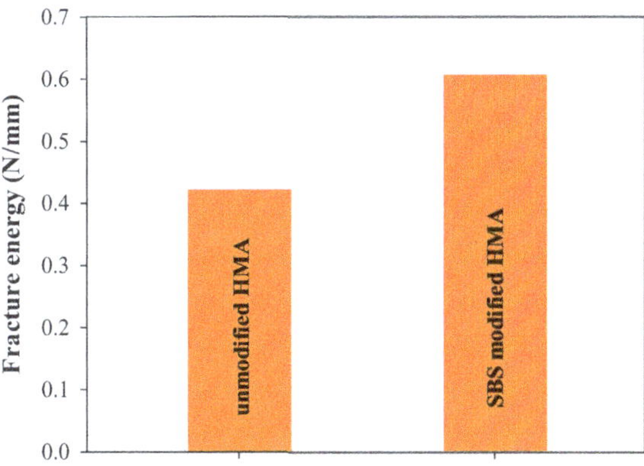

Fig. 3.38 Fracture energy of unmodified and SBS modified HMA concretes [107]

As mentioned earlier, Li and Marasteanu [35] have also investigated the effect of the binder modifier (*i.e.,* SBS) on the fracture energy using the SCB tests at -30, -18, and $-6\,°C$. According to the results, the asphalt mixtures modified with SBS had obviously higher fracture energy than those with plain binder. This improvement in fracture energy was more pronounced at the two lower temperatures, such that the modified mixtures showed a more than 30% increase of fracture energy compared to the unmodified mixture.

In another study, Aliha et al. [108] used five different additives including polyphosphoric acid (PPA), CR, SBS, Sasobit, and anti-stripping agent to investigate their effect on fracture behavior of asphalt concretes. Binder of penetration grade 60/70 was used as a base binder in this study. The specification and preparation procedure of each mixture are briefly described below:

PPA: The modifier PPA was blended with 1% by weight of the base binder at 160 °C for 30 min;
CR: The modifier CR was blended with 15% by weight of the base binder at 175 °C for 120 min using a shear mixer operating at 6000 rpm followed by 60 extra minutes at 150 °C with 1000 rpm;
SBS: The modifier SBS was mixed with 5% by weight of the base binder. The mixing process of the SBS was similar to the one mentioned for the CR;
Sasobit: The modifier Sasobit was blended with 2.5% by weight of the base binder at 130 °C for 10 min using a low shear mixer operating at 300 rpm;
Anti-stripping agent: The anti-stripping agent was mixed with 0.4% by weight of the base binder at 145 °C for 30 min using a low shear mixer operating at 1000 rpm.

Physical properties of the modified binders described above are compared with those of the base binder, as presented in Table 3.6. The letters B, BA, BP, BSa, BSb, and BCR correspond to the base binder, anti-stripping modified binder, Polyphosphoric modified binder, Sosobit modified binder, SBS modified binder and CR modified binder, respectively.

The HMA concretes were prepared by mixing 4.5% of the binders described above and aggregates with the gradation No. 4 (presented in Table 3.2) using a gyratory compactor machine with 90 gyratory rotations to generate an air void content of 3%. It is pointed out that a fixed optimum value of binder (*i.e.,* 4.5% by weight of the HMA mixtures) was used for preparing the whole mixtures to provide a similar condition for all the mixtures. This optimum binder content was obtained from the Marshall mix design of unmodified HMA concrete. Cylindrical samples obtained

Table 3.6 Physical properties of the binders [108]

Property	B	BA	BP	BSa	BSb	BCR
Penetration at 25 °C (0.1 mm)	62	56	48	45	43	44
Ductility at 25 °C (cm)	>100	>100	>100	>100	>100	>100
Softening point (°C)	49	53	61	63	66	67
Performance grade	64–22	64–22	70–22	70–22	70–28	70–28

from the gyratory compactor machine were cut in several steps to produce the SCB specimens with a diameter of 150 mm, thickness of 32 mm, and crack length of 20 mm.

Fracture tests were carried out on the produced SCB specimens under pure mode I, pure mode II, and mixed modes of $M^e = 0.8$ and $M^e = 0.38$ at $-15\,^\circ$C. The crack was located at the middle of the SCB specimen as well as the left bottom support was located at $S_1 = 50$ mm for all the modes of loading, while the right bottom support was located at different positions of $S_2 = 50$, 22, 15 and 9 mm for generating pure mode I, mixed mode of $M^e = 0.8$, mixed mode of $M^e = 0.38$, and pure mode II loadings, respectively. The specimens were cooled in a freezer for 8 h and were then loaded at a displacement rate of 3 mm/min using the three-point bend fixture shown in Fig. 3.8.

According to Fig. 3.39, all the modified HMA mixtures had higher fracture toughness than the unmodified HMA mixture (*i.e.*, B in Fig. 3.39) for all the modes of loading. Furthermore, the HMA mixture containing CR provided the highest fracture toughness among the HMA mixtures considered in this study, and the most improvement in fracture resistance was observed for the CR modified HMA mixture under pure mode II. Swell of binder, due to reaction of CR particles with binder via absorbing aromatic particles of binder, leads to improving the rheological characteristics of binder [109–111]. This issue reduced the lower limit of PG and also increased the higher limit of PG, as given in Table 3.6. On the other hand, capability to absorb high energy and the improved resilient behavior of CR modified binder can enhance its fracture resistance against the crack growth. Another result from Fig. 3.39 is that the least fracture toughness was observed under mixed mode of $M^e = 0.38$ for all the HMA mixtures.

Aliha et al. [75] investigated the effect of SBS, CR, and Lucobit on the mode I fracture toughness of asphalt concrete using the disk-shaped specimen shown in Fig. 3.25.

The SBS and CR were respectively blended with 5 and 15% by weight of the base binder at 180 $^\circ$C for 30 min using a shear mixer operating at 3500 rpm. As such,

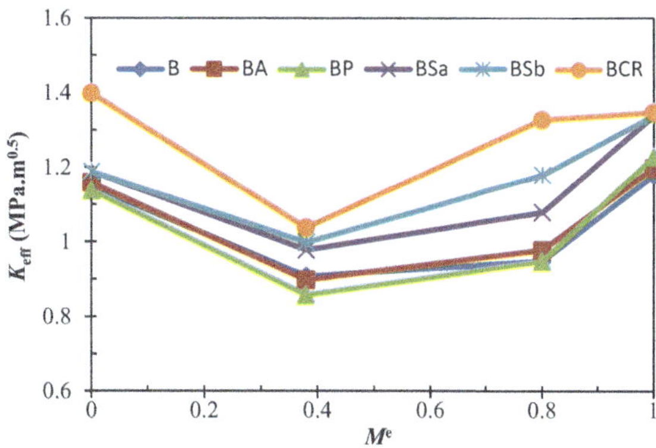

Fig. 3.39 Fracture toughness of HMA mixtures modified by various additives under different modes of loading [108]

4% Lucobit was directly blended with the base binder at 175 °C for 2 min using a screw-type mixer.

The results revealed that all the additives considered in this study had positive influence on enhancing the fracture resistance of asphalt concretes. However, the SBS and Lucobit modified asphalt concretes showed slightly higher fracture toughness than the CR modified one.

3.11 Mixed Mode I/III Fracture Toughness

Previous investigations on the road structures indicated that a cracked asphalt concrete paved on a road experiences all types of fracture modes including mode I, mode II, and mode III depending on the geometry and orientation of crack, position of vehicle wheels relative to crack plane, etc. (see *e.g.*, [112, 113]). These findings directed researchers to investigate fracture behavior of asphalt concretes where they are subjected to mode III loading. Hence, there are some researches in the literature focusing on this issue, which some of them are discussed herein.

Pirmohammad and Kiani [114, 115] investigated mixed mode I/III fracture behavior of HMA mixture using a SCB specimen. The SCB specimen shown in Fig. 3.40

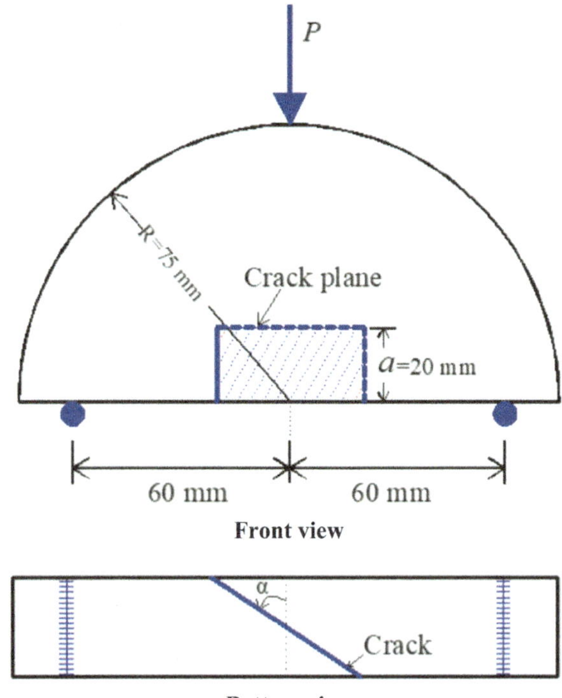

Fig. 3.40 Geometry of the SCB specimen containing a tilted crack [115]

Table 3.7 Finite element results for different crack angles α [114]

α (°)	K_I (MPa $\sqrt{m}$)	K_{III} (MPa$\sqrt{m}$)	Y_I	Y_{III}	M^e
0	0.267	0	5.105	0	1
5	0.267	0.015	5.122	0.282	0.97
10	0.262	0.029	5.026	0.556	0.93
15	0.255	0.043	4.890	0.818	0.89
20	0.245	0.055	4.701	1.063	0.86
25	0.233	0.067	4.464	1.284	0.82
30	0.217	0.076	4.161	1.465	0.79
35	0.201	0.085	3.856	1.637	0.74
40	0.183	0.092	3.496	1.757	0.70
45	0.162	0.096	3.104	1.832	0.66
50	0.140	0.097	2.687	1.863	0.61
55	0.118	0.096	2.251	1.834	0.57
60	0.094	0.091	1.800	1.734	0.51

contained a tilted crack and was loaded by a three-point bend fixture. The parameter α stated the angular position of the crack. According to finite element analysis, pure mode I loading was obtained as the crack was oriented with $\alpha = 0°$. By increasing the crack angle (*i.e.*, α), the mode III was added to the mode I at the crack front, and the SCB specimen was therefore subjected to mixed mode I/III loading. Table 3.7 presents the finite element results for Y_I and Y_{III} which are respectively the mode I and mode III geometry factors. The geometry factors were calculated from the following equations by using the finite element results (*i.e.*, K_I and K_{III}).

$$Y_I = \frac{K_I}{\sqrt{\pi a}} \frac{2Rt}{P} \tag{3.11}$$

$$Y_{III} = \frac{K_{III}}{\sqrt{\pi a}} \frac{2Rt}{P} \tag{3.12}$$

where the values of R (specimen radius), a (crack length), t (thickness), and P (the applied force) are respectively assumed to be 75, 20, 32 mm and 1000 N in the finite element analyses. Meanwhile, the SCB specimen was loaded by a span of 120 mm, and mechanical properties were considered as $E = 12.5$ GPa and $\upsilon = 0.3$.

The mixity parameter M^e presented in Table 3.7, describing the relative proportions of mode I and mode III at the crack tip, is written by:

$$M^e = \frac{2}{\pi} \tan^{-1}\left(\frac{K_I}{K_{III}}\right) \tag{3.13}$$

According to Table 3.7, for the crack position of $\alpha = 0°$, the mixity parameter M^e equals 1, indicating that the specimen is loaded under pure mode I. As the crack position α changed from $0°$ to $60°$, the value of M^e reduced, indicating that proportion of the mode III relative to the mode I increased. It is notable that this specimen was able to produce a limited range of mixed mode I/III loading up to $M^e = 0.5$, and generating further contribution of mode III (*i.e.*, the M^e less than about 0.5) was not feasible due to contacting the crack with the bottom supports.

HMA concrete investigated in this study contained aggregates with gradation No. 4 (as shown in Table 3.2), binder of penetration grade 60/70, and air void of 4%. The SCB specimens made of this mixture had a diameter of 15 cm, thickness of 32 mm, and crack length of 20 mm similar to the values considered in the finite element simulations.

In order to investigate the effect of mode III loading on the fracture behavior of asphalt concrete, the SCB specimens containing an edge crack generated at $\alpha = 0°$ (for pure mode I), $\alpha = 30°$, $45°$, and $60°$ (for mixed mode I/III) were prepared. The fracture tests were performed at a low temperature of $-15\ °C$ using a three-point bend fixture, such that the SCB specimens were placed upon the bottom supports with a span of 120 mm and were then loaded at a displacement rate of 3 mm/min. Afterward, the maximum loads (*i.e.*, P_{cr}) recorded from the tests were put into the following equations to calculate the mode I and mode III critical stress intensity factors.

$$K_{If} = Y_I \frac{P_{cr}}{2Rt} \sqrt{\pi a} \qquad\qquad (3.14)$$

$$K_{IIIf} = Y_{III} \frac{P_{cr}}{2Rt} \sqrt{\pi a} \qquad\qquad (3.15)$$

The effective critical SIF (*i.e.*, K_{eff}), indicating the fracture strength of materials subjected to any combination of modes I and III, can be calculated from the following equation:

$$K_{eff} = \sqrt{K_{If}^2 + K_{IIIf}^2} \qquad\qquad (3.16)$$

It is reminded that Eqs. 3.14–3.16 are valid for the cases that fracture takes place in accordance with the linear elastic fracture mechanics (LEFM) approach, in other words when brittle fracture occurs during the experiments.

Figure 3.41 displays the fracture surface of the SCB specimens under different modes of loading. Three various regions were observed on the fracture surface: (i) the first region was related to the initial crack, (ii) the second region, corresponding to the fracture surface near the top fixture, appeared as a flat surface. In this region, the crack has propagated based on mode I fracture because the fracture surface was approximately flat and along the applied load, and (iii) the third region corresponded to the fracture surface between the initial crack and the second region. Indeed, this region connected the initial crack front (which was inclined to the mixed mode I/III

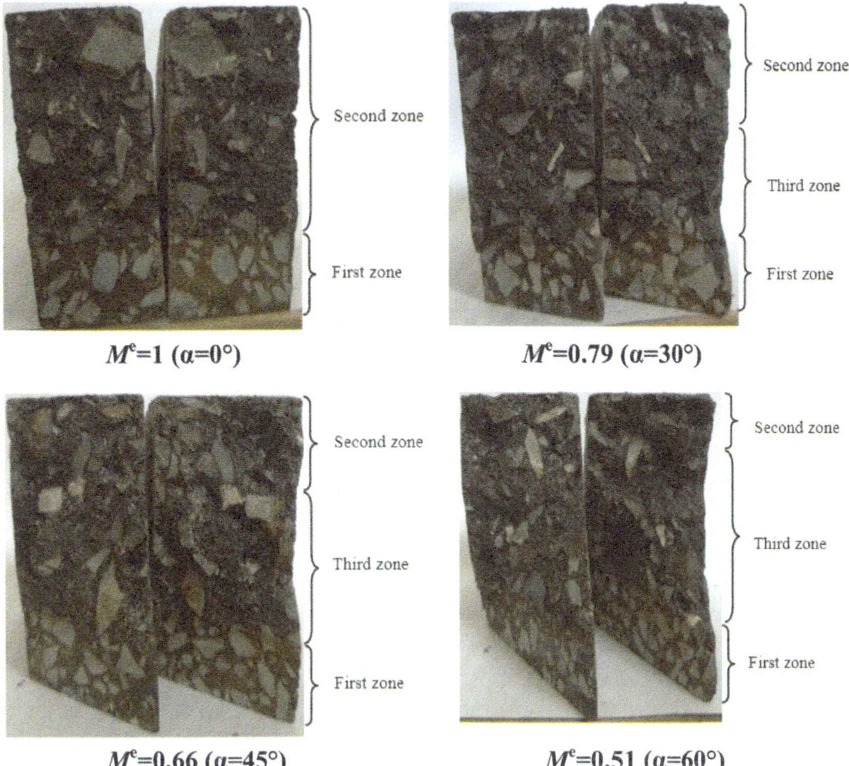

Fig. 3.41 Fracture surface of the SCB specimen under different modes of loading [114]

cases) to the second flat region. Hence, this region gradually twisted as the crack propagated toward the top fixture. Furthermore, as the proportion of mode III at the crack front of the SCB specimen enhanced (*i.e.,* α increased), the third region became larger.

Based on the results given in Fig. 3.42, the fracture resistance of HMA mixture reduced as the proportion of mode III increased. Hence, the presence of mode III deformation at crack tip can weaken HMA mixture against the fracture.

In other studies, Pirmohammad and Bayat [116–118] investigated the fracture behavior of HMA concretes under mixed mode I/III loading at three different temperatures (-5, -20, and -30 °C). They used two types of HMA mixtures containing aggregates with gradation No. 4 (shown in Table 3.2) and an air void of 4%. The first mixture, called control HMA concrete, contained base binder of penetration grade 60/70, and the second one, called modified HMA concrete, was modified by 15% CR (by weight of the base binder).

A disk bend specimen shown in Fig. 3.43 was also used for simulating different modes of loading. This specimen had a radius of $R = 100$ mm, thickness of $t = 40$ mm, and crack length of $a = 16$ mm. For loading the specimen, it was put upon

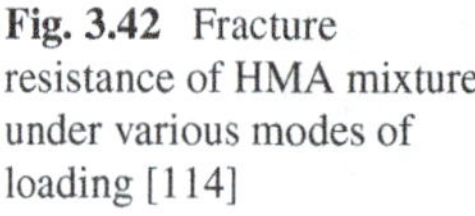

Fig. 3.42 Fracture resistance of HMA mixture under various modes of loading [114]

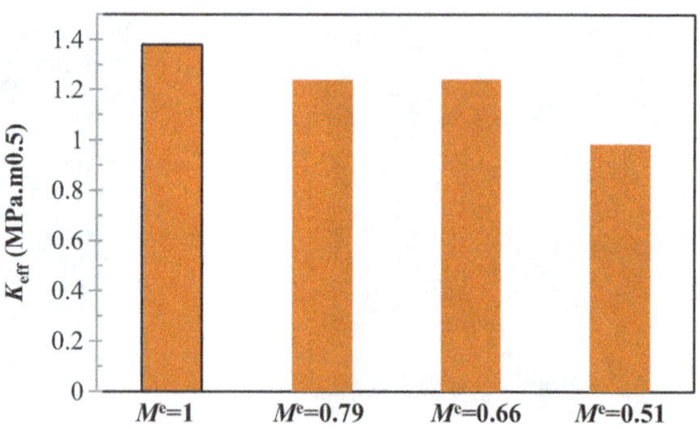

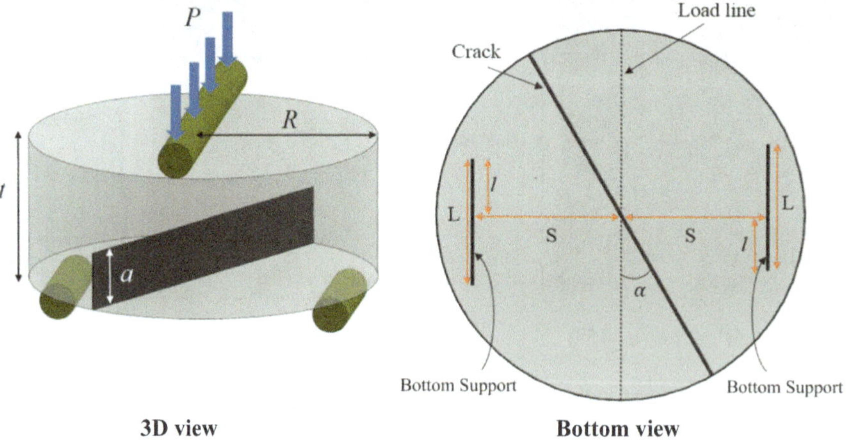

3D view   **Bottom view**

Fig. 3.43 Geometry of disk bend specimen [118]

the bottom supports of length $L = 34$ mm with span of $2S = 80$, and the force P was then applied at the top and middle of the specimen. The bottom supports had an offset of $l = 15$ mm relative to the middle line of the specimen (see Fig. 3.43).

As shown in Fig. 3.43, the bottom supports did not cover throughout the specimen and had a length of $L = 34$ mm. Finite element results showed that pure mode III was achieved at high values of α (*i.e.*, 63°–65° [116]), depending on the sizes of specimen and crack. This caused the bottom supports to locate at regions very near to the specimen edges. But, inappropriate failure took place during the fracture experiments on asphalt concrete, *i.e.*, failure occurred at the region around the bottom supports due to high stress concentration (see Fig. 3.44). Hence, for avoiding this drawback, the bottom supports should be moved away from the specimen edges. This led to another problem, contacting the bottom supports with the crack. In order to sort this problem out, the length of bottom supports should be shortened. For generating full range of mixed mode loading, *i.e.*, pure mode I, mixed mode I/III, and pure mode

Fig. 3.44 Inappropriate failure occurred around the bottom supports [118]

III, the specimen shown in Fig. 3.43 was finally suggested for the porous materials like asphalt mixtures.

The results of finite element analyses given in Table 3.8 showed that the disk bend specimen was subjected to pure mode I loading as the crack was located at $\alpha = 0°$. By rotating the crack (*i.e.*, by increasing α), in addition to mode I deformation, the

Table 3.8 Finite element results for different crack angles α [117]

α (°)	K_I (MPa $\sqrt{m}$)	K_III (MPa $\sqrt{m}$)	Y_I	Y_III
0	0.208	0	0.309	0
5	0.206	0.0091	0.307	0.0142
10	0.201	0.0186	0.299	0.0276
15	0.192	0.0279	0.286	0.0415
20	0.178	0.0372	0.265	0.0554
25	0.163	0.0453	0.243	0.0673
30	0.145	0.0520	0.215	0.0773
35	0.124	0.0570	0.184	0.0847
40	0.102	0.0599	0.152	0.0890
45	0.0807	0.0622	0.120	0.0924
50	0.0584	0.0623	0.0868	0.0927
55	0.0357	0.0609	0.0531	0.0905
60	0.0133	0.0579	0.0198	0.0860
63	0	0.0553	0	0.0821

mode III deformation was also present at the crack tip; therefore, the specimen was subjected to mixed mode I/III loading. The contribution of mode III increased when the crack rotated further (*i.e.*, α increased), and the specimen was finally subjected to pure mode III as the crack oriented at $\alpha = 63°$. The geometry factors (*i.e.*, Y_I and Y_{III}) for the disk bend specimen can be calculated from the following equations by using the finite element results (*i.e.*, K_I and K_{III} given in Table 3.8).

$$Y_I = \frac{K_I}{\sqrt{\pi a}}\frac{Rt^2}{6PS} \tag{3.17}$$

$$Y_{III} = \frac{K_{III}}{\sqrt{\pi a}}\frac{Rt^2}{6PS} \tag{3.18}$$

It is useful to note that the mechanical properties were considered as $E = 12.5$ GPa and $\upsilon = 0.3$, and the applied force P was assumed to be 1000 N in the finite element analyses.

Fracture experiments were conducted on the two control and modified HMA mixtures under pure mode I ($\alpha = 0°$), pure mode III ($\alpha = 63°$), and two intermediate mixed modes of I/III ($\alpha = 30°$ and $\alpha = 45°$) at three different temperatures ($-5, -20,$ and -30 °C). The disk bend specimens with the dimensions mentioned above were initially cooled in a freezer for 4 h and were then loaded at a constant displacement rate of 3 mm/min. The critical forces P_{cr} recorded from the tests were put into the following equations to calculate the critical SIFs:

$$K_{If} = Y_I\frac{6SP_{cr}}{Rt^2}\sqrt{\pi a} \tag{3.19}$$

$$K_{IIIf} = Y_{III}\frac{6SP_{cr}}{Rt^2}\sqrt{\pi a} \tag{3.20}$$

The values of critical SIFs (K_{If} and K_{IIIf}) were placed into Eq. 3.16 to calculate the fracture strength (*i.e.*, K_{eff}) of the HMA mixtures at different temperatures, which are plotted in Fig. 3.45.

Figure 3.45 shows the fracture resistance of the control and modified HMA concretes at different temperatures and under various modes of loading. According to this figure, by decreasing the temperature, the fracture resistance of both the control and modified asphalt concretes initially increased and then decreased due to the DTC damage occurring in the mixture, as described earlier. This trend was observed for all the modes of loading (*i.e.*, pure mode I, pure mode III, and mixed modes of I/III). Furthermore, the HMA mixture modified by CR demonstrated higher fracture resistance than the control HMA concrete under any mode of loading. The improvement in fracture resistance by application of CR was dependent on the mode of loading.

According to Fig. 3.45, by increasing the contribution of mode III relative to mode I from pure mode I to pure mode III, the fracture resistance of both asphalt concretes decreased gradually. Aliha et al. [119, 120] reported similar results when considering mixed mode I/III fracture behavior of asphalt concrete. According to their results,

Fig. 3.45 Fracture strength of HMA mixtures at different temperatures under **a** pure mode I ($\alpha = 0°$), **b** mixed mode I/III ($\alpha = 30°$), **c** mixed mode I/III ($\alpha = 45°$), and **d** pure mode III ($\alpha = 63°$) [117]

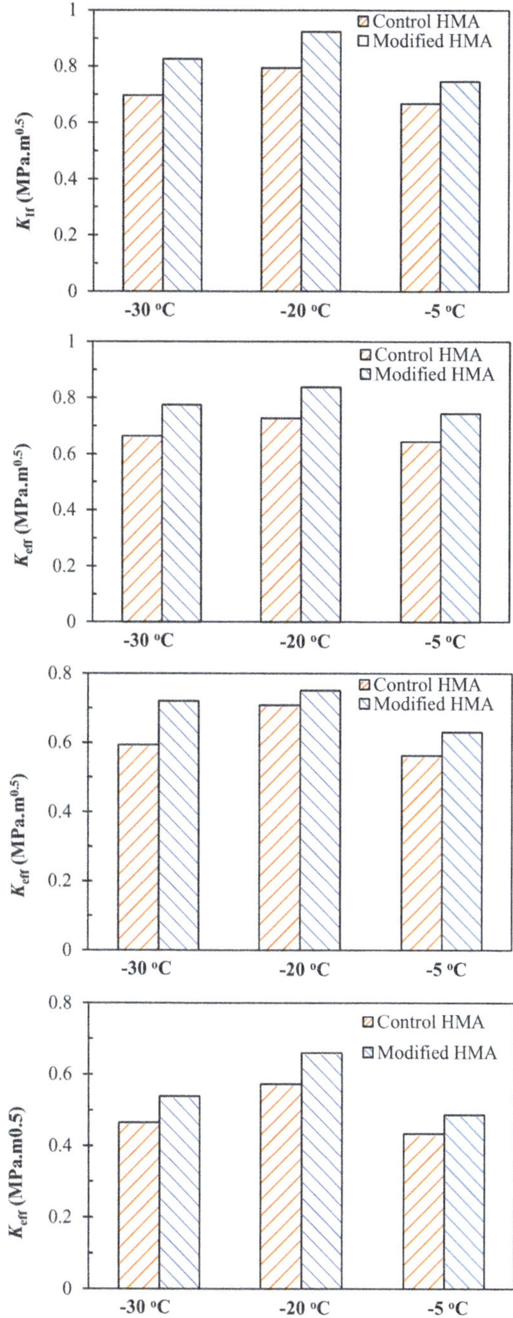

the mode III fracture toughness of asphalt concrete was less than its mode I fracture toughness, implying that asphalt pavements are more vulnerable to fracture under mode III deformation than when they are loaded under pure mode I.

It is also pointed out that the trend observed for the mixed mode I/III fracture resistance of asphalt concretes is contrary to that observed for the case of mixed mode I/II loading, which by enhancing the proportion of mode II relative to mode I, the fracture resistance of asphalt concretes initially decreases and then increases, as discussed earlier.

The nature of mixed mode I/III fracture is sophisticated to some extent in comparison with the mixed mode I/II. This is because the crack extension under mixed mode I/III loading leads to multiple fracture planes intersecting the crack front. Hence, the fracture surfaces were often nonplanar, and the crack twisted along its propagation path, as shown in Fig. 3.46. Indeed, higher mode III contribution along the crack front resulted in larger twisting in the fracture trajectory.

3.12 Effect of Specimen Size and Geometry on Fracture Test Results

According to the previous investigations performed on asphalt mixtures [5, 121–123] and other materials like rocks [124–128], polymers [129], etc., both the geometry and size of specimen can influence the magnitude of fracture indicators (*i.e.,* fracture toughness, fracture energy, etc.). Some of the researches performed concerning the effect of specimen type and size on asphalt concretes and other materials are reviewed below.

Artamendi and Khalid [5] measured the fracture energy of asphalt mixtures using the SENB and SCB tests. They used the SENB specimens with dimensions of 305 mm in length, 65 mm in height, and 50 mm in thickness for conducting pure mode I and mixed mode I/II experiments. A notch with 19.5 mm in length was fabricated in the SENB specimens. The notch was located in the centerline of the SENB specimens for pure mode I experiments, while for the mixed mode I/II experiments, it was created in the SENB specimen with an offset of 48.8 mm from the centerline. In addition, the span was assumed to be 244 mm in the experiments.

For the SCB tests, diameter, thickness, and crack length were taken 153 mm, 65 mm, and 23 mm, respectively. For pure mode I, the notch was located at the centerline with an angle of 0°, while for the mixed mode I/II loading, the notch was generated at the centerline, but with an angle of 45°, with respect to the direction of the load applied to the specimen at the top of the SCB specimen.

Two types of asphalt mixtures (*i.e.,* dense bitumen macadam (DBM) and stone mastic asphalt (SMA)) were used in this study. Details of these mixtures together with the procedure of fracture experiments have been given in Chap. 5.

For the mode I experiments, the temperatures were selected as -10, 0, and 10 °C, while the mixed mode I/II experiments were performed only at a temperature of 0 °C.

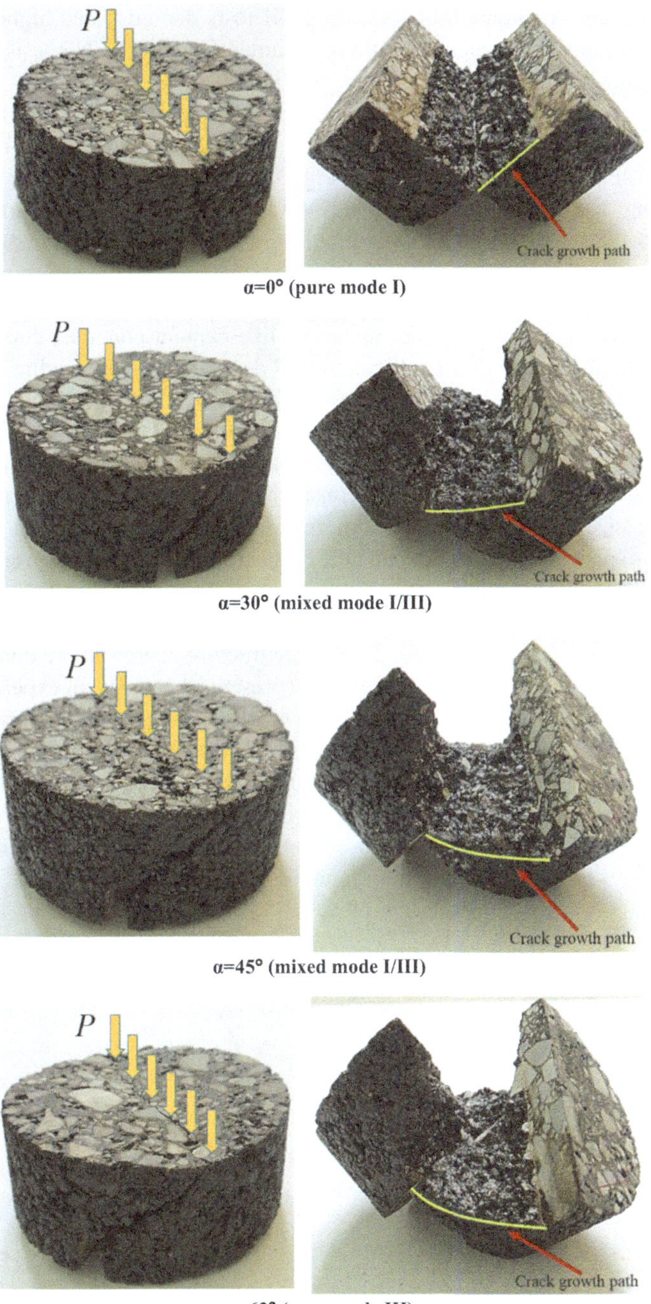

Fig. 3.46 Fracture trajectory in HMA mixtures under different modes of loading [118]

According to Fig. 3.14 and Table 5.5, the SCB tests demonstrated higher fracture energies (*i.e.,* more than twice) than those obtained from the SENB tests for all the test temperatures and loading conditions mentioned above.

In another study, Kim et al. [122] investigated the size effect of DC(T) specimen on the fracture energy of asphalt concrete. For this purpose, the DC(T) specimen with four different dimensions shown in Fig. 3.47 was produced.

Limestone aggregates with 9.5 mm NMAS and asphalt binder with PG64-22 were employed to prepare asphalt mixture. Furthermore, asphalt binder and air void contents of mixture were 5.25% and 6.5%, respectively.

Experiments were conducted at $-10\ ^\circ\mathrm{C}$ to decrease the viscous response of the asphalt concrete. Contrary to other materials like cement concrete, ceramics, and glass, the fracture energy of asphalt concrete is dependent on the loading rate; thus, different loading rates were considered for different sized specimens to ensure that the fracture energy comparison between the different specimen sizes was valid. The value of *CMOD* rate was 1 mm/min for the standard DC(T) specimen as per ASTM E399; therefore, the *CMOD* rates were considered as 0.667, 2, and 3 mm/min for the specimen sizes of 100, 300, and 450 mm, respectively.

In order to perform experiment, the DC(T) specimen was put inside a temperature chamber set at $-10\ ^\circ\mathrm{C}$ and was loaded in tension through the loading holes with a constant *CMOD* rate mentioned above. The load–*CMOD* curve recorded from the DC(T) test was finally used to calculate the fracture energy (more details on the DC(T) test can be found in Chap. 5). Table 3.9 presents the results of experiments for different dimensions of the DC(T) specimen, in which increase in dimensions of the DC(T) specimen resulted in enhancement of the fracture energy. This difference in the values of fracture energy was attributed to the viscus behavior of asphalt concrete.

Molenaar et al. [10] investigated the size effect of SCB specimens on the fracture toughness of asphalt concrete. They used the SCB specimens with three different diameters (*i.e.,* 100, 150, and 220 mm) and thicknesses (*i.e.,* 25, 50, and 75 mm) to

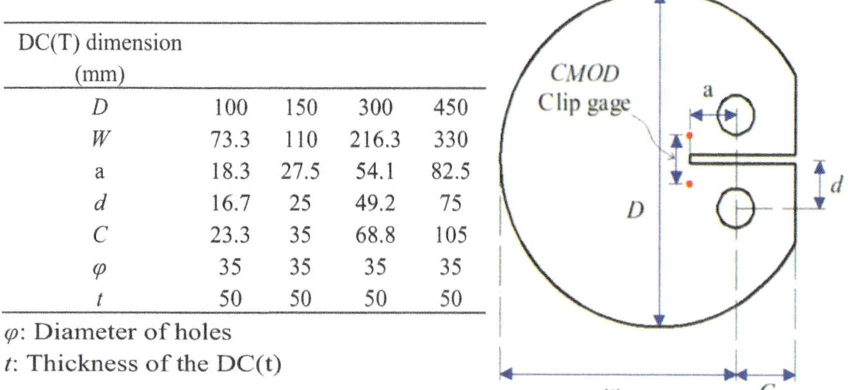

DC(T) dimension (mm)				
D	100	150	300	450
W	73.3	110	216.3	330
a	18.3	27.5	54.1	82.5
d	16.7	25	49.2	75
C	23.3	35	68.8	105
φ	35	35	35	35
t	50	50	50	50

φ: Diameter of holes

t: Thickness of the DC(t)

Fig. 3.47 Geometry and dimensions of DC(T) specimen used in [122]

Table 3.9 Fracture energy obtained from the DC(T) tests with different dimensions [122]

Diameter (mm)	Number of samples	Mean fracture energy (N/m)	Standard deviation	Coefficient of variation
100	8	404.2	72.7	18
150	8	410.8	62.5	15.2
300	8	529.5	67.3	12.7
450	5	623.6	94.7	15.2

obtain the fracture toughness of asphalt concrete at different temperatures (*i.e.,* − 10, 0, 15, and 25 °C). On the basis of results, the fracture toughness was found to be independent of the SCB diameter (ranging from 100 to 220 mm) and thickness (ranging from 25 to 75 mm) at the temperature of 0 °C or lower. However, the fracture toughness decreased as the SCB diameter increased at the temperature of 15 °C.

Ayatollahi and Akbardoost [125] investigated the size effect of center-cracked circular disk (CCCD) specimen subjected to pure mode I loading (see Fig. 3.48). Dimensions of this specimen with different sizes are presented in Table 3.10. Material of the specimens was a type of rock (*i.e.,* white marble). The specimens having different sizes were compressively loaded using the UTM with a cross-head speed of 0.5 mm/min. Table 3.10 shows the mode I fracture toughness of rock specimens with different sizes. According to the results, the fracture toughness increased as the size of the CCCD specimen enhanced.

Fig. 3.48 Center-cracked circular disk specimen subjected to mode I loading [125]

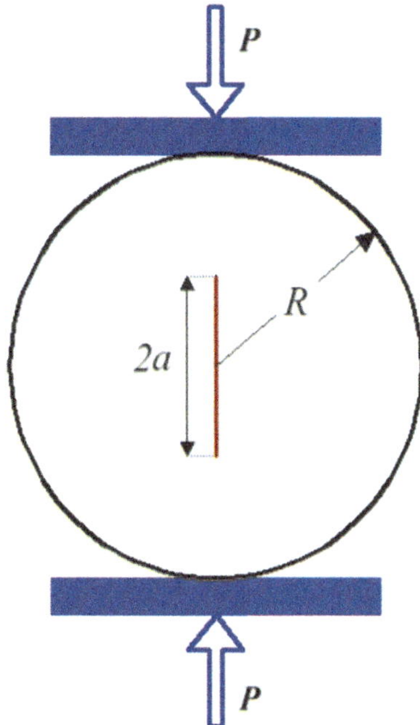

Table 3.10 Dimensions of CCCD specimen with different sizes and the test results [125]

Specimen size	R (mm)	a (mm)	t (mm)	K_{If} (MPa m$^{0.5}$)
Size 1	30	9	17.5	0.523
Size 2	50	15	17.5	0.696
Size 3	75	22.5	17.5	0.742
Size 4	95	28.5	18	0.905

Fig. 3.49 SENB specimen subjected to pure mode I loading [130]

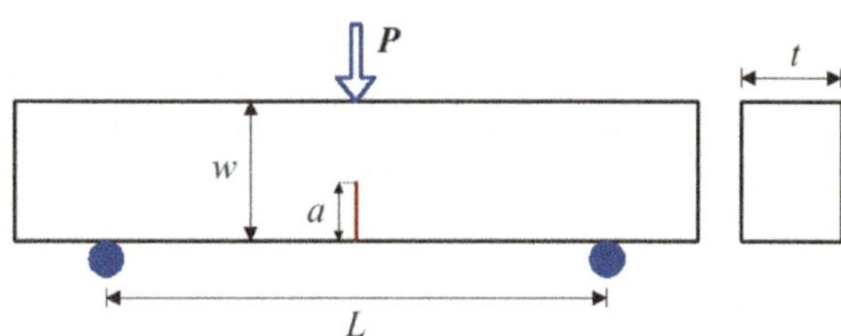

Table 3.11 Dimensions of SENB specimen with different sizes and the test results [127, 130]

Dimensions of SENB ($L \times w \times t$)	K_{If} (MPa m$^{0.5}$)
$408 \times 102 \times 13$	0.774
$204 \times 51 \times 13$	0.651
$100 \times 25 \times 13$	0.532
$52 \times 13 \times 13$	0.436

Bazant et al. [130] performed experiments on the SENB specimens made of Indiana limestone to explore the size effect. Figure 3.49 displays the geometry of SENB specimen subjected to pure mode I loading, and its sizes are also given in Table 3.11. The crack length was selected as $a/w = 0.4$ in the specimen preparation. According to Table 3.11, increase in the size of SENB specimen led to enhancement of the fracture toughness.

It is worth mentioning that Bazant et al. [131] performed similar SENB tests made of concrete, and the results exhibited that increase of the size of SENB specimen resulted in nonlinear reduction of the fracture toughness.

In another research, Ayatollahi and Akbardoost investigated the size effect of SCB specimen (shown in Fig. 3.50c) made of rock with different dimensions and subjected to pure mode I and pure mode II loadings (see Table 3.12). The experiments were performed with a span ratio of $S/R = 0.5$ and constant displacement rate of 0.5 mm/min. Based on the results given in Table 3.12, both the values of mode I and mode II fracture toughness increased as the SCB size increased.

Aliha et al. [129] measured the mixed mode I/II fracture toughness of polymethylmethacrylate (PMMA) using four different types of specimens namely symmetric edge cracked triangular specimen (SECT), asymmetric edge cracked triangular specimen (AECT), symmetric edge cracked semicircular specimen (SECS), and asymmetric edge cracked semicircular specimen (AECS) shown in Fig. 3.50. Different

Table 3.12 Dimensions of SCB specimen with different dimensions and the test results [126]

Specimen size ($R \times t \times a$) (mm)	α (°)	K_{If} (MPa m$^{0.5}$)	K_{IIf} (MPa m$^{0.5}$)
$25 \times 26 \times 12.5$	0	0.983	–
	41	–	0.553
$50 \times 26 \times 25$	0	1.317	
	41		0.713
$95 \times 26 \times 47.5$	0	1.493	
	41		0.821
$190 \times 26 \times 95$	0	1.65	
	41		0.914

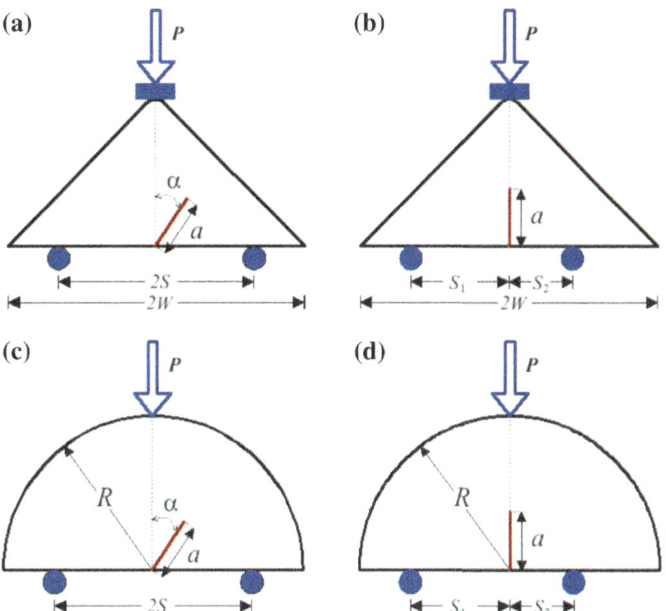

Fig. 3.50 **a** SECT, **b** AECT, **c** SECS, **d** AECS specimen [129]

mode mixities ranging from pure mode I to pure mode II can be obtained using these specimens by changing the crack inclination angle (α) or the support location (*i.e.*, S_2). Symmetry in geometry (*i.e.*, α) and loading condition ($S_1 = S_2$) resulted in generation of pure mode I. Mode II was also observed at the crack front by increasing the value of α or decreasing the value of S_2 till pure mode II was achieved at a special value of α or S_2. Critical SIFs of the abovementioned specimens can be calculated from the following equations:

For triangular specimens:

$$K_{\mathrm{If}} = Y_{\mathrm{I}}\frac{P_{\mathrm{cr}}}{2Wt}\sqrt{\pi a} \qquad (3.21)$$

$$K_{\mathrm{IIf}} = Y_{\mathrm{II}}\frac{P_{\mathrm{cr}}}{2Wt}\sqrt{\pi a} \qquad (3.22)$$

For semicircular specimens:

$$K_{\mathrm{If}} = Y_{\mathrm{I}}\frac{P_{\mathrm{cr}}}{2Rt}\sqrt{\pi a} \qquad (3.23)$$

$$K_{\mathrm{IIf}} = Y_{\mathrm{II}}\frac{P_{\mathrm{cr}}}{2Rt}\sqrt{\pi a} \qquad (3.24)$$

Fracture toughness (*i.e.*, K_{eff}) of any material can be measured by replacing the values of critical SIFs (*i.e.*, K_{If} and K_{IIf}) into Eq. 3.16. Y_{I} and Y_{II} are the mode I and mode II geometry factors which have been documented in [129] for different values of α, a, S, S_1, S_2, W, R, and t.

In order to compare the results of fracture toughness using the mentioned four specimens, the values of a, R, and W were assumed to be the same in the experimental program performed by Aliha et al. [129]. According to the results, although all the test specimens were loaded using three-point bending fixture, completely different fracture toughnesses were obtained for different mixed mode I/II loading conditions. Hence, different geometries or loading conditions produced different values of fracture toughness. It is also noted that the discrepancy of fracture results became more pronounced by increasing the proportion of mode II at the crack front. Furthermore, the specimens containing vertical crack showed significantly higher fracture toughness than those containing inclined crack. The results of this study indicated that fracture behavior of real structures subjected to any arbitrary loading can be influenced by the geometry and loading conditions.

3.13 Fracture Process Zone in Asphalt Concrete

The size of fracture process zone (FPZ) is an important parameter to predict failure and to choose the geometry and dimensions of the test specimen because the ligament (*i.e.*, the zone ahead of crack tip) should be large enough to encompass the FPZ. In addition, the FPZ can be helpful in the numerical methods of fracture using phenomenological and micromechanical models like the CZM [132–135]. Thus, characterization of the process zone size at a microlevel can be highly useful to calibrate the phenomenological model and therefore to increase the precision of numerical simulation.

Acoustic emission (AE) methods are widely employed to determine the microscopic fracture processes and to assess the damage growth in asphalt mixtures. The AE method has been successfully used in the past years to understand the relation between the microstructural events and the macroscopic performance of the asphalt mixtures (see *e.g.,* [136–138]). Furthermore, the AE method has been also used by many researchers to investigate the FPZ in the fracture tests for different materials [135].

Li and Marasteanu [135] captured the evolution and size of FPZ around the notch tip of SCB specimens made of different asphalt mixtures, representing a combination of different factors like aggregate type, binder type, binder content, loading rate, and air voids, at three low temperatures (*i.e.,* -6, -18, and -30 °C).

They used three binders of PG58-28, PG64-28, and PG64-28+SBS and two different types of aggregates including granite and limestone with NMAS of 12.5 mm to prepare the mixtures. In addition, two levels of air voids (*i.e.,* 4 and 7%) were selected to study the effect of air voids on the size of FPZ. Two levels of binder content including the design value and the design value plus 0.5% were employed in the preparation of mixtures. The specifications of the mixtures considered in [135] are presented in Table 3.13. The first part of the mixture ID presents the high limit of the binder PG, and the second part refers to the type of binder (*i.e.,* P: plain binder or S: SBS modified binder). The third and fourth parts correspond to the level of air voids (*i.e.,* 4% or 7%) and the type of aggregates (G: granite or L: limestone), respectively. In addition, design binder contents of 6% and 6.9% were respectively used for the granite and limestone mixtures.

The SCB specimen of 25 mm thickness and 150 mm diameter was used in this study. A notch with 15 mm in length and 2 mm in width was generated in the SCB specimen. In order to study the effect of the notch length on the size of FPZ, the SCB specimens with two additional notch lengths of 5 mm and 30 mm were produced for the mixture 64:S:4:L.

The experiments were performed in an environmental chamber. The *LLD* was measured using a vertically mounted extensometer; one end was mounted on a button

Table 3.13 Asphalt concrete specifications [135]

Mixture ID	Air voids (%)		Aggregate type		Binder content	
	4	7	Granite	Limestone	Design	Design +0.5%
58:P:4:G	√		√		√	
58:P:4:G+0.5	√		√			√
58:P:4:L	√			√	√	
58:P:4:L+0.5	√			√		√
58:P:7:L		√		√	√	
64:P:4:L	√			√	√	
64:S:4:L	√			√	√	

that was permanently fixed on a specially made frame, and the other end was attached to a metal button glued to the SCB specimen (see Fig. 5.3). The experiments were conducted at a *CMOD* rate of 0.03 mm/min. Before loading the specimens, they were maintained in the environmental chamber at the test temperature for two hours to avoid any temperature gradient within the specimens.

In the fracture tests using the SCB specimen, the AE event signals were recorded by four DAQ cards. Each card included two independent channels which acquired AE signals detected by eight piezoelectric sensors. Four sensors were mounted on each side of the SCB specimens.

The AE energy for each event was computed as the integral of the signal amplitude square during the event duration. The events with very little energy may have very little effect on the fracture process. The FPZ of asphalt mixtures subjected to the testing conditions can be also estimated by the distribution of the AE energy for the AE events taking place prior to the peak load.

The effect of different factors including temperature, aggregate, air voids, binder content, loading rate, and notch length on the FPZ size has been investigated in [135].

Since asphalt concrete is a temperature-dependent material, it is of interest to study the effect of temperature on the FPZ size. Figure 3.51 shows the FPZ for the

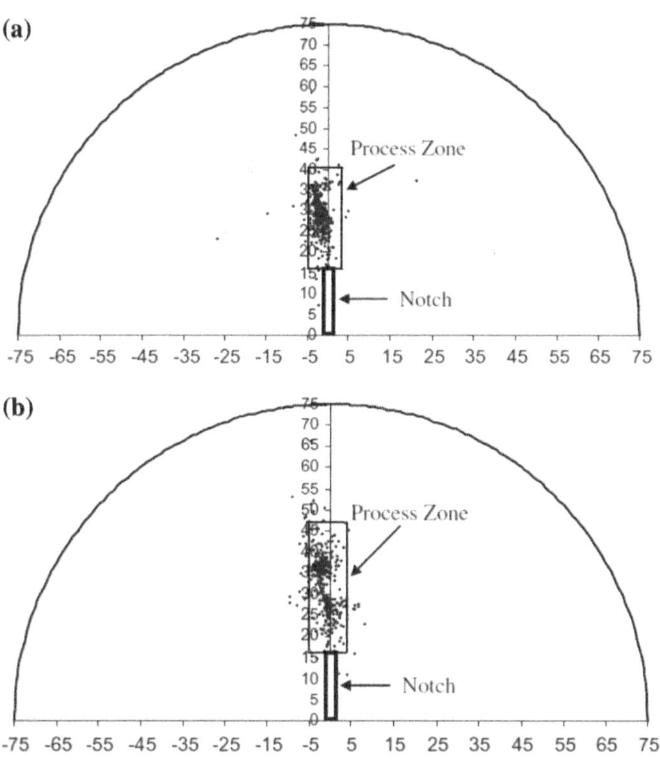

Fig. 3.51 Fracture process zone at **a** −18 °C, **b** −30 °C [135]

mixture of 58:P:4:G+0.5 at the test temperatures of -18 and $-30\ °C$. It is noticed that very limited events were observed at the temperature of $-6\ °C$, and therefore, the FPZ was not obtained. Comparing the results shown in Fig. 3.51 indicated that there was no obvious difference in the width of FPZ but the FPZ was about 8 mm longer at the temperature of $-30\ °C$ than at the temperature of $-18\ °C$.

They measured the FPZ of the mixtures of 58:P:4:G +0.5 and 58:P:4:L +0.5 at $-30\ °C$ to investigate the effect of aggregate type on the FPZ size. The lengths of the FPZ obtained for the mentioned two mixtures were very close; however, the mixture containing limestone aggregates had 5 mm larger width of FPZ than that containing granite aggregates. The narrower FPZ for the mixture containing granite signified that the micro-cracking took place in a smaller area ahead of the crack tip prior to propagating the crack.

Comparison of the results obtained for the mixtures containing different air voids of 4 and 7% showed that the mixture with 7% air voids (*i.e.*, 58:P:7:L at $-30\ °C$) had a much bigger FPZ than that with 4% air voids (*i.e.*, 58:P:4:L at $-30\ °C$), signifying that the micro-cracking took place in a much bigger area ahead of the crack tip in the mixture with higher air void content prior to the peak load. In other words, the mixture with 7% air voids was more damaged at the microscale level because of having higher air voids.

According to the results, no obvious difference in the size of FPZ was observed between the mixtures with two different binder contents (*i.e.*, the mixtures of 58:P:4:G and 58:P:4:G+0.5 at $-30\ °C$).

They also investigated the effect of loading rate on the FPZ size of asphalt mixture. For this purpose, the experiments were performed on the mixture of 64:P:4:L at the temperature of $-30\ °C$ with three different *CMOD* rates of 0.009, 0.03, and 0.3 mm/min. No obvious difference in the size of FPZ was observed; however, the limited data documented in [135] seemed to confirm the conclusion that the FPZ was a property of the material.

Comparison of the results for the notch lengths of 5, 15, and 30 mm (on the SCB specimens made of 64:S:4:L at $-30\ °C$) indicated no obvious difference in the width of the FPZ size; however, shorter notch demonstrated longer FPZ (*i.e.*, the length of FPZ was 47, 40, and 25 mm for the notch lengths of 5, 15, and 30 mm, respectively).

3.14 Summary

In this chapter, the fracture behavior of HMA concretes was reviewed. Different test specimens including the SENB, DC(T), SCB, etc. have been used by researchers to measure the fracture toughness of HMA concretes. However, the SCB specimen was mostly employed for this purpose because of the following reasons:

(i) HMA samples are usually cylindrical, achievable from the gyratory compactor machine, field coring, and Marshal compactor.

(ii)   Conventional three-point bend fixture is easily used for loading the SCB specimen.
(iii) Full mode mixities (*i.e.*, pure mode I, pure mode II, and any mixed mode I/II) can be generated at the crack tip of the SCB specimen.

The procedure of performing the fracture tests using the mentioned specimens was also described in this chapter.

In the subsequent sections, the effects of different parameters including aggregate type, aggregate gradation, air void content, binder, and temperature on the fracture toughness of HMA mixtures were discussed.

Furthermore, the effects of nanomaterials, fibers, and additives on the fracture toughness of HMA concretes were reviewed. The results showed that these materials can improve pure mode I, pure mode II, and mixed mode I/II fracture toughness of HMA mixtures significantly.

The fracture behavior of HMA concretes subjected to mixed mode I/III and pure mode III loadings was also evaluated. The results revealed that the presence of mode III can worsen the fracture resistance remarkably.

The effect of specimen geometry and size on the fracture behavior of asphalt concrete and some other materials were also reviewed. The results indicated that both the geometry and size of specimen can influence the fracture characteristics of materials, and this issue should be also regarded in the design of real road structures.

The effect of different parameters such as temperature, aggregate, air voids, binder content, loading rate, and notch length on the FPZ size was finally discussed.

References

1. Mobasher B, Mamlouk MS, Lin HM (1997) Evaluation of crack propagation properties of asphalt mixtures. J Transp Eng 123(5):405–413
2. Kim KW, Doh YS, Lim S (1999) Mode I reflection cracking resistance of strengthened asphalt concretes. Constr Build Mater 13(5):243–251
3. Kim KW, El Hussein M (1997) Variation of fracture toughness of asphalt concrete under low temperatures. Constr Build Mater 11(7–8):403–411
4. Kim KW, Kweon SJ, Doh YS, Park T-S (2003) Fracture toughness of polymer-modified asphalt concrete at low temperatures. Can J Civ Eng 30(2):406–413
5. Artamendi I, Khalid HA (2006) A comparison between beam and semi-circular bending fracture tests for asphalt. Road Mater Pavement Des 7(sup1):163–180
6. Mamlouk M, Mobasher B (2004) Cracking resistance of asphalt rubber mix versus hot-mix asphalt. Road Mater Pavement Des 5(4):435–451
7. Wagoner MP, Buttlar WG, Paulino GH (2005) Development of a single-edge notched beam test for asphalt concrete mixtures. J Test Eval 33(6):452–460
8. Wagoner M, Buttlar WG, Paulino G (2005) Disk-shaped compact tension test for asphalt concrete fracture. Exp Mech 45(3):270–277
9. Wagoner M, Buttlar W, Paulino G, Blankenship P (1929) Investigation of the fracture resistance of hot-mix asphalt concrete using a disk-shaped compact tension test. Transp Res Rec J Transp Res Board 2005:183–192
10. Molenaar J, Liu X, Molenaar A (2003) Resistance to crack-growth and fracture of asphalt mixture. In: 6th International RILEM symposium, Zurich, Switzerland, 14–16 Apr 2003

11. Ayatollahi MR, Pirmohammad S (2013) Temperature effects on brittle fracture in cracked asphalt concretes. Struct Eng Mech 45(1):19–32
12. Aliha MM, Behbahani H, Fazaeli H, Rezaifar M (2014) Study of characteristic specification on mixed mode fracture toughness of asphalt mixtures. Constr Build Mater 54:623–635
13. Pirmohammad S, Khoramishad H, Ayatollahi M (2015) Effects of asphalt concrete characteristics on cohesive zone model parameters of hot mix asphalt mixtures. Can J Civ Eng 43(3):226–232
14. Ayatollahi M, Aliha M, Hassani M (2006) Mixed mode brittle fracture in PMMA—an experimental study using SCB specimens. Mater Sci Eng, A 417(1–2):348–356
15. Ayatollahi M, Aliha M, Saghafi H (2011) An improved semi-circular bend specimen for investigating mixed mode brittle fracture. Eng Fract Mech 78(1):110–123
16. Kim H, Wagoner MP, Buttlar WG (2009) Micromechanical fracture modeling of asphalt concrete using a single-edge notched beam test. Mater Struct 42(5):677
17. Xie T, Qiu YJ, Jiang ZZ, Al CF (2006) Study on compound type crack propagation behavior of asphalt concrete. Key Eng Mater 759–762
18. Gao H, Yang X, Zhang C (2015) Experimental and numerical analysis of three-point bending fracture of pre-notched asphalt mixture beam. Constr Build Mater 90:1–10
19. Song SH, Paulino GH, Buttlar WG (2006) A bilinear cohesive zone model tailored for fracture of asphalt concrete considering viscoelastic bulk material. Eng Fract Mech 73(18):2829–2848
20. Braham A, Buttlar W (2009) Mode II cracking in asphalt concrete. Adv Test Charact Bituminous Mater 2:699–706
21. Im S, Ban H, Kim YR (2014) Characterization of mode-I and mode-II fracture properties of fine aggregate matrix using a semicircular specimen geometry. Constr Build Mater 52:413–421
22. Ayatollahi M, Aliha M (2007) Fracture toughness study for a brittle rock subjected to mixed mode I/II loading. Int J Rock Mech Min Sci 44(4):617–624
23. Aliha M, Ayatollahi M (2010) Geometry effects on fracture behaviour of polymethyl methacrylate. Mater Sci Eng, A 527(3):526–530
24. Ayatollahi M, Aliha M (2006) On determination of mode II fracture toughness using semi-circular bend specimen. Int J Solids Struct 43(17):5217–5227
25. Akbardoost J, Ayatollahi M, Aliha M, Pavier M, Smith D (2014) Size-dependent fracture behavior of Guiting limestone under mixed mode loading. Int J Rock Mech Min Sci 71:369–380
26. Aliha M, Ayatollahi M (2013) Two-parameter fracture analysis of SCB rock specimen under mixed mode loading. Eng Fract Mech 103:115–123
27. Aliha M, Ayatollahi M, Akbardoost J (2012) Typical upper bound–lower bound mixed mode fracture resistance envelopes for rock material. Rock Mech Rock Eng 45(1):65–74
28. Ayatollahi M, Aliha M (2007) Wide range data for crack tip parameters in two disc-type specimens under mixed mode loading. Comput Mater Sci 38(4):660–670
29. Ameri M, Mansourian A, Pirmohammad S, Aliha M, Ayatollahi M (2012) Mixed mode fracture resistance of asphalt concrete mixtures. Eng Fract Mech 93:153–167
30. Pirmohammad S, Ayatollahi M (2014) Fracture resistance of asphalt concrete under different loading modes and temperature conditions. Constr Build Mater 53:235–242
31. Pirmohammad S, Kiani A (2016) Effect of temperature variations on fracture resistance of HMA mixtures under different loading modes. Mater Struct 49(9):3773–3784
32. Braham AF, Buttlar WG, Marasteanu MO (2007) Effect of binder type, aggregate, and mixture composition on fracture energy of hot-mix asphalt in cold climates. Transp Res Rec 2001(1):102–109
33. Anderson DA, Goetz WH (1973) Mechanical behavior and reinforcement of mineral filler-asphalt mixtures
34. Buttlar WG, Bozkurt D, Al-Khateeb GG, Waldhoff AS (1999) Understanding asphalt mastic behavior through micromechanics. Transp Res Rec 1681(1):157–169
35. Li XJ, Marasteanu M (2010) Using semi circular bending test to evaluate low temperature fracture resistance for asphalt concrete. Exp Mech 50(7):867–876

36. Li X, Marasteanu MO, Kvasnak A, Bausano J, Williams RC, Worel B (2010) Factors study in low-temperature fracture resistance of asphalt concrete. J Mater Civ Eng 22(2):145–152
37. Pirmohammad S, Ayatollahi M (2015) Asphalt concrete resistance against fracture at low temperatures under different modes of loading. Cold Reg Sci Technol 110:149–159
38. Christensen DW, Bonaquist R, Anderson DA, Gokhale S (2004) Indirect tension strength as a simple performance test, performance tests for asphalt mixes, p 44
39. Birgisson B, Montepara A, Napier J, Romeo E, Tebaldi G (2005) Evaluation of aggregate size-dependent of asphalt mixtures in cracking behavior. In: 3rd international SIIV congress, Bari, Italy, 22–24 Sept 2005
40. Li X, Braham AF, Marasteanu MO, Buttlar WG, Williams RC (2008) Effect of factors affecting fracture energy of asphalt concrete at low temperature. Road Mater Pavement Des 9(sup1):397–416
41. Bhurke A, Shin E, Drzal L (1997) Fracture morphology and fracture toughness measurement of polymer-modified asphalt concrete. Transp Res Rec J Transp Res Board (1590):23–33
42. Champion L, Gerard J, Planche J, Martin D, Anderson D (1999) Evaluation of the low-temperature fracture properties of modified binders. In: Relationship with their micromorphology, workshop briefing, performance related properties for bituminous binders, Eurobitume Workshop, Luxembourg
43. Champion-Lapalu L, Planche J, Martin D, Anderson D, Gerard J (2000) Low-temperature rheological and fracture properties of polymer-modified Bitumens. In: 2nd Euroasphalt & Eurobitume Congress. Barcelona, Citeseer, p 122
44. Polacco G, Stastna J, Biondi D, Zanzotto L (2006) Relation between polymer architecture and nonlinear viscoelastic behavior of modified asphalts. Curr Opin Colloid Interface Sci 11(4):230–245
45. Togunde OP, Hesp S (2008) Low temperature investigations on asphalt binder performance—a case study on highway 417 trial sections
46. Hussein HME, Halim AAE (1993) Differential thermal expansion and contraction: a mechanistic approach to adhesion in asphalt concrete. Can J Civ Eng 20(3):366–373
47. Kim KW, El Hussein H (1995) Effect of differential thermal contraction on fracture toughness of asphalt materials at low temperatures (with discussion). J Assoc Asphalt Paving Technol 64
48. Mansourian A, Razmi A, Razavi M (2016) Evaluation of fracture resistance of warm mix asphalt containing jute fibers. Constr Build Mater 117:37–46
49. Pirmohammad S, Kiani A (2016) Impact of temperature cycling on fracture resistance of asphalt concretes. Comput Concr 17(4):541–551
50. Erdogan F, Sih G (1963) On the crack extension in plates under plane loading and transverse shear. J Basic Eng 85(4):519–525
51. Pirmohammad S, Shabani H (2019) Mixed mode I/II fracture strength of modified HMA concretes subjected to different temperature conditions. J Test Eval 47(5):3355–3371
52. Ye C, Chen H (2009) Study on road performance of nano SiO_2 and nano TiO_2 modified asphalt. New Build Mater 6:027
53. Sadeghnejad M, Arabani M, Shakeri V, Mirabdolazimi S (2013) Experimental evaluation of fatigue life of asphalt mixture modified with nano zinc oxide. In: 7th congress on civil engineering, Zahedan, IR Iran
54. Jahromi SG, Khodaii A (2009) Effects of nanoclay on rheological properties of bitumen binder. Constr Build Mater 23(8):2894–2904
55. Khattak MJ, Khattab A, Rizvi HR, Zhang P (2012) The impact of carbon nano-fiber modification on asphalt binder rheology. Constr Build Mater 30:257–264
56. Shafabakhsh G, Ani OJ (2015) Experimental investigation of effect of Nano TiO2/SiO2 modified bitumen on the rutting and fatigue performance of asphalt mixtures containing steel slag aggregates. Constr Build Mater 98:692–702
57. Ashish PK, Singh D, Bohm S (2016) Evaluation of rutting, fatigue and moisture damage performance of nanoclay modified asphalt binder. Constr Build Mater 113:341–350

58. Zhang HL, Su MM, Zhao SF, Zhang YP, Zhang ZP (2016) High and low temperature properties of nano-particles/polymer modified asphalt. Constr Build Mater 114:323–332

59. Amin I, El-Badawy SM, Breakah T, Ibrahim MH (2016) Laboratory evaluation of asphalt binder modified with carbon nanotubes for Egyptian climate. Constr Build Mater 121:361–372

60. Arabani M, Faramarzi M (2015) Characterization of CNTs-modified HMA's mechanical properties. Constr Build Mater 83:207–215

61. Ziari H, Farahani H, Goli A, Sadeghpour Galooyak S (2014) The investigation of the impact of carbon nano tube on bitumen and HMA performance. Petrol Sci Technol 32(17):2102–2108

62. Xiao F, Amirkhanian AN, Amirkhanian SN (2010) Influence of carbon nanoparticles on the rheological characteristics of short-term aged asphalt binders. J Mater Civ Eng 23(4):423–431

63. Kordi Z, Shafabakhsh G (2017) Evaluating mechanical properties of stone mastic asphalt modified with Nano Fe2O3. Constr Build Mater 134:530–539

64. Pirmohammad S, Majd-Shokorlou Y, Amani B (2020) Experimental investigation of fracture properties of asphalt mixtures modified with Nano Fe_2O_3 and carbon nanotubes. Road Mater Pavement Des (in press)

65. Timm D, Birgisson B, Newcomb D (1998) Development of mechanistic-empirical pavement design in Minnesota. Transp Res Rec J Transp Res Board (1629):181–188

66. Pérez-Jiménez F, Botella R, Moon KH, Marasteanu M (2013) Effect of load application rate and temperature on the fracture energy of asphalt mixtures, Fénix and semi-circular bending tests. Constr Build Mater 48:1067–1071

67. Treacy MJ, Ebbesen T, Gibson J (1996) Exceptionally high Young's modulus observed for individual carbon nanotubes. Nature 381(6584):678

68. Bai J, Allaoui A (2003) Effect of the length and the aggregate size of MWNTs on the improvement efficiency of the mechanical and electrical properties of nanocomposites—experimental investigation. Compos A Appl Sci Manuf 34(8):689–694

69. Chong KP, Garboczi EJ (2002) Smart and designer structural material systems. Prog Struct Mat Eng 4(4):417–430

70. Makar J, Beaudoin J (2004) Carbon nanotubes and their application in the construction industry. Roy Soc Chem 292:331–342

71. Pirmohammad S, Majd-Shokorlou Y, Amani B (2019) Fracture resistance of HMA mixtures modified with nanoclay and Nano-Al_2O_3. J Test Eval 47(5):3289–3308

72. You Z, Mills-Beale J, Foley JM, Roy S, Odegard GM, Dai Q, Goh SW (2011) Nanoclay-modified asphalt materials: preparation and characterization. Constr Build Mater 25(2):1072–1078

73. Al-Mansob RA, Ismail A, Rahmat RAO, Borhan MN, Alsharef JM, Albrka SI, Karim MR (2017) The performance of epoxidised natural rubber modified asphalt using nano-alumina as additive. Constr Build Mater 155:680–687

74. Ameri M, Nowbakht S, Molayem M, Aliha M (2016) Investigation of fatigue and fracture properties of asphalt mixtures modified with carbon nanotubes. Fatigue Fract Eng Mater Struct 39(7):896–906

75. Aliha M, Sarbijan M, Bahmani A (2017) Fracture toughness determination of modified HMA mixtures with two novel disc shape configurations. Constr Build Mater 155:789–799

76. Aliha M, Razmi A, Mansourian A (2017) The influence of natural and synthetic fibers on low temperature mixed mode I + II fracture behavior of warm mix asphalt (WMA) materials. Eng Fract Mech 182:322–336

77. Tapkın S, Uşar Ü, Tuncan A, Tuncan M (2009) Repeated creep behavior of polypropylene fiber-reinforced bituminous mixtures. J Transp Eng 135(4):240–249

78. Wu S, Ye Q, Li N (2008) Investigation of rheological and fatigue properties of asphalt mixtures containing polyester fibers. Constr Build Mater 22(10):2111–2115

79. Dehghan Z, Modarres A (2017) Evaluating the fatigue properties of hot mix asphalt reinforced by recycled PET fibers using 4-point bending test. Constr Build Mater 139:384–393

80. Serin S, Morova N, Saltan M, Terzi S (2012) Investigation of usability of steel fibers in asphalt concrete mixtures. Constr Build Mater 36:238–244

81. Moghadas Nejad F, Vadood M, Baeetabar S (2014) Investigating the mechanical properties of carbon fibre-reinforced asphalt concrete. Road Mater Pavement Des 15(2):465–475
82. Wang X, Wu R, Zhang L (2019) Development and performance evaluation of epoxy asphalt concrete modified with glass fibre. Road Mater Pavement Des 20(3):715–726
83. Klinsky L, Kaloush K, Faria V, Bardini V (2018) Performance characteristics of fiber modified hot mix asphalt. Constr Build Mater 176:747–752
84. Apostolidis P, Liu X, Daniel GC, Erkens S, Scarpas T (2019) Effect of synthetic fibres on fracture performance of asphalt mortar. Road Mater Pavement Des 1–14
85. Sani MA, Latib AA, Ng CP, Yusof MA, Ahmad N, Rani MAM (2011) Properties of coir fibre and kenaf fibre modified asphalt mixes. In: Proceedings of the Eastern Asia society for transportation studies the 9th international conference of Eastern Asia society for transportation studies, 2011, Eastern Asia Society for Transportation Studies, pp 258–258
86. Mughal MA, Hainin MR, Jaya RP, Yunus MF, Rahman M, Idham MK, Warid MNM (2018) Stabilizing asphalt concrete using kenaf fibers. Adv Sci Lett 24(6):3963–3967
87. Hainin M, Idham M, Yaro N, Hussein S, Warid M, Mohamed A, Naqibah S, Ramadhansyah P (2018) Performance of hot mix asphalt mixture incorporating kenaf fibre. In: IOP conference series: earth and environmental science. IOP Publishing, p 012092
88. Pirmohammad S, Majd-Shokorlou Y, Amani B (2020) Influence of natural fibers (kenaf and goat wool) on mixed mode I/II fracture strength of asphalt mixtures. Constr Build Mater 239 (in press)
89. Pirmohammad S, Amani B, Majd-Shokorlou Y (2020) The effect of basalt fibers on fracture toughness of asphalt mixture. Fatigue Fract Eng Mater Struct (in press)
90. Pirmohammad S, Majd-Shokorlou Y, Amani B (2020) Laboratory investigations on fracture toughness of asphalt concretes reinforced with carbon and kenaf fibers. Eng Fract Mech 226 (in press)
91. Arabani M, Shabani A (2019) Evaluation of the ceramic fiber modified asphalt binder. Constr Build Mater 205:377–386
92. Abtahi SM, Sheikhzadeh M, Hejazi SM (2010) Fiber-reinforced asphalt-concrete–a review. Constr Build Mater 24(6):871–877
93. Chen H, Xu Q, Chen S, Zhang Z (2009) Evaluation and design of fiber-reinforced asphalt mixtures. Mater Des 30(7):2595–2603
94. Lavasani M, Namin ML, Fartash H (2015) Experimental investigation on mineral and organic fibers effect on resilient modulus and dynamic creep of stone matrix asphalt and continuous graded mixtures in three temperature levels. Constr Build Mater 95:232–242
95. Committee A (2008) State of the art report on fiber reinforced concrete-ACI 544.1 R-96 (Reapproved 2002). ACI manual of concrete practice, Part 6
96. Elsaid A, Dawood M, Seracino R, Bobko C (2011) Mechanical properties of kenaf fiber reinforced concrete. Constr Build Mater 25(4):1991–2001
97. Qin X, Shen A, Guo Y, Li Z, Lv Z (2018) Characterization of asphalt mastics reinforced with basalt fibers. Constr Build Mater 159:508–516
98. Qi Y, Fan W, Nan G (2017) Free-standing, binder-free polyacrylonitrile/asphalt derived porous carbon fiber–A high capacity anode material for sodium-ion batteries. Mater Lett 189:206–209
99. Mahrez A, Karim MR, Bt Katman HY (2005) Fatigue and deformation properties of glass fiber reinforced bituminous mixes. J Eastern Asia Soc Transp Stud 6:997–1007
100. Stempihar JJ, Souliman MI, Kaloush KE (2012) Fiber-reinforced asphalt concrete as sustainable paving material for airfields. Transp Res Rec 2266(1):60–68
101. Fu S-Y, Lauke B, Mäder E, Yue C-Y, Hu X (2000) Tensile properties of short-glass-fiber- and short-carbon-fiber-reinforced polypropylene composites. Compos A Appl Sci Manuf 31(10):1117–1125
102. Fazaeli H, Amini AA, Nejad FM, Behbahani H (2016) Rheological properties of bitumen modified with a combination of FT paraffin wax (sasobit®) and other additives. J Civ Eng Manag 22(2):135–145
103. Al-Hadidy A (2001) Influence of polyethylene and sulfur wastes on characteristics of asphalt paving materials, M.Sc. thesis, College of Engineering, Civil Engineering Department, Al

104. Punith V, Veeraragavan A (2007) Behavior of asphalt concrete mixtures with reclaimed polyethylene as additive. J Mater Civ Eng 19(6):500–507
105. Casey D, McNally C, Gibney A, Gilchrist MD (2008) Development of a recycled polymer modified binder for use in stone mastic asphalt. Resour Conserv Recycl 52(10):1167–1174
106. Bhurke AS, Shin EE, Drzal LT (1997) Fracture morphology and fracture toughness measurement of polymer-modified asphalt concrete. Transp Res Rec 1590(1):23–33
107. Pirmohammad S (2013) Study of mixed mode I/II fracture behavior of asphaltic materials at low temperatures, Ph.D. dissertation, Mechanical Engineering, Iran University of Science and Technology
108. Aliha M, Fazaeli H, Aghajani S, Nejad FM (2015) Effect of temperature and air void on mixed mode fracture toughness of modified asphalt mixtures. Constr Build Mater 95:545–555
109. Yadollahi G, Mollahosseini HS (2011) Improving the performance of crumb rubber bitumen by means of poly phosphoric acid (PPA) and Vestenamer additives. Constr Build Mater 25(7):3108–3116
110. Edwards Y, Tasdemir Y, Isacsson U (2006) Rheological effects of commercial waxes and polyphosphoric acid in bitumen 160/220—low temperature performance. Fuel 85(7–8):989–997
111. Edwards Y, Tasdemir Y, Isacsson U (2007) Rheological effects of commercial waxes and polyphosphoric acid in bitumen 160/220–high and medium temperature performance. Constr Build Mater 21(10):1899–1908
112. Ameri M, Mansourian A, Khavas MH, Aliha M, Ayatollahi M (2011) Cracked asphalt pavement under traffic loading–A 3D finite element analysis. Eng Fract Mech 78(8):1817–1826
113. Ayatollahi MR, Pirmohammad S, Sedighiani K (2014) Three-dimensional finite element modeling of a transverse top-down crack in asphalt concrete. Comput Concr 13(4):569–585
114. Pirmohammad S, Kiani A (2016) Study on fracture behavior of HMA mixtures under mixed mode I/III loading. Eng Fract Mech 153:80–90
115. Pirmohammad S, Kiani A (2018) Numerical analysis of a new specimen for conducting fracture experiments under mixed mode I/III loading. J Solids Fluids Mech 8(3):247–259
116. Pirmohammad S, Bayat A (2016) Finite element analysis of a new specimen for conducting fracture tests under mixed mode I/III loading. J Stress Anal 1(1):33–41
117. Pirmohammad S, Bayat A (2017) Fracture resistance of HMA mixtures under mixed mode I/III loading at different subzero temperatures. Int J Solids Struct 120:268–277
118. Pirmohammad S, Bayat A (2016) Characterizing mixed mode I/III fracture toughness of asphalt concrete using asymmetric disc bend (ADB) specimen. Constr Build Mater 120:571–580
119. Aliha M, Bahmani A, Akhondi S (2016) A novel test specimen for investigating the mixed mode I + III fracture toughness of hot mix asphalt composites–Experimental and theoretical study. Int J Solids Struct 90:167–177
120. Aliha M, Bahmani A, Akhondi S (2015) Determination of mode III fracture toughness for different materials using a new designed test configuration. Mater Des 86:863–871
121. Zegeye E, Le J-L, Turos M, Marasteanu M (2012) Investigation of size effect in asphalt mixture fracture testing at low temperature. Road Mater Pavement Des 13(sup1):88–101
122. Kim H, Wagoner MP, Buttlar WG (2009) Numerical fracture analysis on the specimen size dependency of asphalt concrete using a cohesive softening model. Constr Build Mater 23(5):2112–2120
123. Molenaar J, Molenaar A (2000) Fracture toughness of asphalt in the semi-circular bend test. In: 2nd Eurasphalt and Eurobitume Congress, Barcelona, Spain, 20–22 Sept 2000
124. Aliha M, Sistaninia M, Smith D, Pavier M, Ayatollahi M (2012) Geometry effects and statistical analysis of mode I fracture in guiting limestone. Int J Rock Mech Min Sci 51:128–135
125. Ayatollahi M, Akbardoost J (2014) Size and geometry effects on rock fracture toughness: mode I fracture. Rock Mech Rock Eng 47(2):677–687
126. Ayatollahi M, Akbardoost J (2013) Size effects in mode II brittle fracture of rocks. Eng Fract Mech 112:165–180

127. Ayatollahi M, Akbardoost J (2012) Size effects on fracture toughness of quasi-brittle materials–A new approach. Eng Fract Mech 92:89–100

128. Khoramishad H, Akbardoost J, Ayatollahi M (2014) Size effects on parameters of cohesive zone model in mode I fracture of limestone. Int J Damage Mech 23(4):588–605

129. Aliha M, Bahmani A, Akhondi S (2016) Mixed mode fracture toughness testing of PMMA with different three-point bend type specimens. Eur J Mech-A/Solids 58:148–162

130. Bažant Z, Gettu R, Kazemi M (1991) Identification of nonlinear fracture properties from size effect tests and structural analysis based on geometry-dependent R-curves. Int J Rock Mech Min Sci Geomech Abst 43–51

131. Bažant ZP, Kim J-K, Pfeiffer PA (1986) Nonlinear fracture properties from size effect tests. J Struct Eng 112(2):289–307

132. Soares JB, Freitas F, Allen DH (1832) Crack modeling of asphaltic mixtures considering heterogeneity of the material. Transp Res Rec 2003:113–120

133. Paulino GH, Song SH, Buttlar WG (2004) Cohesive zone modeling of fracture in asphalt concrete. In: Proceedings of the 5th international RILEM conference–cracking in pavements: mitigation, risk assessment, and preservation, Limoges, France, pp 63–70

134. Li X, Marasteanu MO (2005) Cohesive modeling of fracture in asphalt mixtures at low temperatures. Int J Fract 136(1–4):285–308

135. Li X, Marasteanu M (2010) The fracture process zone in asphalt mixture at low temperature. Eng Fract Mech 77(7):1175–1190

136. Li X, Marasteanu MO (2006) Investigation of low temperature cracking in asphalt mixtures by acoustic emission. Road Mater Pavement Des 7(4):491–512

137. Wendling L, Xolin E, Gimenez D, Reynaud P, De La Roche C, Chevalier J, Fantozzi G (2004) Characterisation of crack propagation in bituminous mixtures. In: Fifth international RILEM conference on cracking in pavements, Limoges, France, pp 639–646

138. Hesp SA, Smith BJ, Hoare TR (2001) Effect of the filler particle size on the low and high temperature performance in asphalt mastic and concrete. J Assoc Asphalt Paving Technol 70:492–508

Chapter 4
Fracture Behavior of WMA Concretes

Abstract This chapter deals with the fracture behavior of warm mix asphalt (WMA) concretes. The effects of different parameters including mode of loading (i.e., pure mode I, pure mode II, and mixed mode I/II), temperature, crumb rubber, fiber (e.g., jute, kenaf, etc.), and aggregate type on fracture resistance of WMA mixtures are reviewed. Although WMA concretes are prepared at temperatures 20–55 °C lower than HMA concretes, fracture toughness of WMA is as high as that of HMA, and even some additives used in WMA mixtures provide higher fracture toughness as compared with HMA mixtures.

4.1 Introduction

Warm mix asphalt (WMA) concretes are suitable alternatives for pavement materials; because, they are prepared at temperatures 20–55 °C lower than conventional hot mix asphalt (HMA) concretes. Decrease in greenhouse gases, saving in fuel consumption, easy compaction of mixtures, and capability for mixing and paving at low temperatures are among the main benefits of WMA technology.

At least 20 additives or processes have been introduced over the past years to prepare WMA concretes. Generally, WMA technologies can be classified into three groups [1]:

(i)   Organic additives: this category includes fatty acid amide or wax additives that liquefy below the mixing temperature of WMA;
(ii) Chemical additives: these products are able to influence the surface bonding between binder and aggregate;
(iii) Foaming processes and additives: water is used in this category to foam binder and entirely coat aggregates at a decreased preparation temperature.

WMA mixtures are relatively new, and have been developed over the last decade, so concerns still exist on the performance of WMA mixtures. However, several attempts have been made in recent years to study some basic characteristics (e.g., rutting, fatigue behavior, etc.) of WMA concretes. Researchers have also employed crumb rubber and different fibers such as kenaf, jute, FORTA, and basalt to improve the fracture strength of WMA mixtures. This chapter deals with mechanical behavior

© Springer Nature Switzerland AG 2020

S. Pirmohammad and M. R. Ayatollahi, *Fracture Behavior of Asphalt Materials*, Structural
Integrity 14, https://doi.org/10.1007/978-3-030-39974-0_4

of WMA with particular emphasis on its fracture properties under a combination of tensile and shear loads.

Since basic information on the fracture tests (i.e., test specimens, test set-up, test procedures, modes of loading, etc.) have been discussed in Chap. 3, so the study of Chap. 3 is recommended to the readers before studying the current chapter.

4.2   Mixed Mode I/II Fracture Resistance of WMA Concretes

Although WMA concretes are relatively new technology amongst the pavement systems; however, many attempts have been made in the past years to investigate performance of WMA concretes [2]. For example, Frigio et al. [2] studied the effect of aging on WMA concrete. Hasan et al. [3] studied the effect of moisture on the performance of WMA concrete, and used hydrated lime as an anti-stripping agent for improving its moisture susceptibility. According to Tafti et al. [4], binder consumption reduces by utilizing siliceous aggregates instead of limestone aggregates during preparation of WMA concrete. An investigation performed by Su et al. [5] shows that WMA concrete containing a coarse gradation provides a comparable performance in comparison to HMA concretes, and WMA concrete is, therefore, recommended for use in the rehabilitation of airport overlays. Based on Fakhri et al. [6], the fatigue life of WMA mixtures is longer than HMA ones at low strain levels whilst the fatigue life of WMA concretes is comparable with that of HMA ones at high strain levels. Das et al. [7] investigated the cracking behavior of WMA mixtures containing wax by performing indirect tensile tests at low temperatures. The results indicated that the wax has a negative influence on the cracking behavior of WMA mixtures.

Several investigations (e.g., [8–11]) can be also found on fracture behavior of WMA concretes, which are discussed below.

4.2.1   Effect of Temperature

The temperature dependence of asphalt concretes has been the subject of several studies. For example, Pirmohammad and Khanpour [10] investigated the effect of temperature variations on mixed mode I/II fracture behavior of WMA concretes. The SCB specimen (shown in Fig. 3.4a) used in this research had a diameter of 150 mm, thickness of 32 mm, and crack length of 20 mm. The WMA concretes studied in this research contained a mixture of base binder with 60/70 penetration grade and 0.3% Sasobit. To prepare this mixture, the Sasobit was blended with the base binder at 135 °C for 10 min using a low shear mixer operating at 300 rpm. Meanwhile, aggregates with gradation No. 4 presented in Table 3.2 were used in the preparation of WMA concretes. In addition, the WMA concretes contained an air void of 4%.

Table 4.1 Numerical results for the SCB specimen shown in Fig. 3.4a [10]

Loading mode	M^e	S_1, S_2 & L (mm)	K_I $(MPa\sqrt{m})$	K_{II} $(MPa\sqrt{m})$	Y_I	Y_{II}
Pure mode I	1	$(S_1, S_2) = (60, 60)$ & $L = 0$	0.267	0	5.105	0
Mixed mode I/II	0.8	$(S_1, S_2) = (50, 20)$ & $L = -2$	0.086	0.029	1.655	0.546
Mixed mode I/II	0.5	$(S_1, S_2) = (50, 20)$ & $L = 5$	0.061	0.059	1.171	1.131
Mixed mode I/II	0.2	$(S_1, S_2) = (50, 20)$ & $L = 11$	0.031	0.094	0.599	1.792
Pure mode II	0	$(S_1, S_2) = (50, 20)$ & $L = 16$	0	0.120	0	2.298

It is reminded that the melting point of Sasobit is about 100 °C, and it is solvable in binder at this temperature. In addition, Sasobit does not segregate on storage at temperatures above 120 °C [8].

The crack location (i.e., L) and bottom support locations (i.e., S_1 and S_2) in the SCB specimen for simulating different modes of loading together with the corresponding values of the geometry factors (i.e., Y_I and Y_{II}) are given in Table 4.1. The mixity parameter M^e expressing the relative contributions of mode I and mode II can be calculated using Eq. (3.1). In addition, geometry factors (i.e., Y_I and Y_{II}) can be calculated from Eqs. (3.2) and (3.3). More information on this issue can be found in Chap. 3.

In this research, fracture experiments were performed at the following environmental conditions:

(i)   The specimens were conditioned at a constant temperature (CT) of -15 °C for 4 h;

(ii) The specimens were exposed to a variable temperature (VT) condition. Figure 3.17 plots profile of the VT condition, in which the temperature decreases from 25 to -30 °C, and then increases up to 25 °C. The rate of temperature change was 6 h per half cycle. Meanwhile, the specimens experienced the VT condition for 7 days before performing the tests.

In order to compare the results of the VT condition with those of the CT one, the specimens conditioned at the VT were finally kept at the temperature of -15 °C for 4 h similar to the CT condition.

After exposing the specimens to the abovementioned temperature conditions (i.e., CT or VT), they were put upon the bottom supports of the universal test machine (UTM) adjusted at the proper locations (i.e., S_1 and S_2 given in Table 4.1), and the top fixture was finally moved downwards at a displacement rate of 3 mm/min to load the specimens. It is pointed out that the experiments were conducted under pure mode I, pure mode II, and three different mixed modes of $M^e = 0.8$, $M^e = 0.5$, and $M^e = 0.2$.

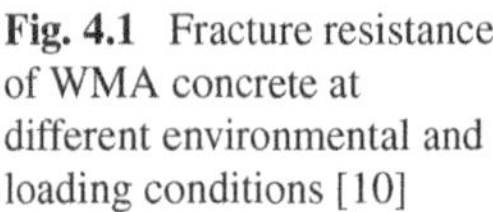

Fig. 4.1 Fracture resistance of WMA concrete at different environmental and loading conditions [10]

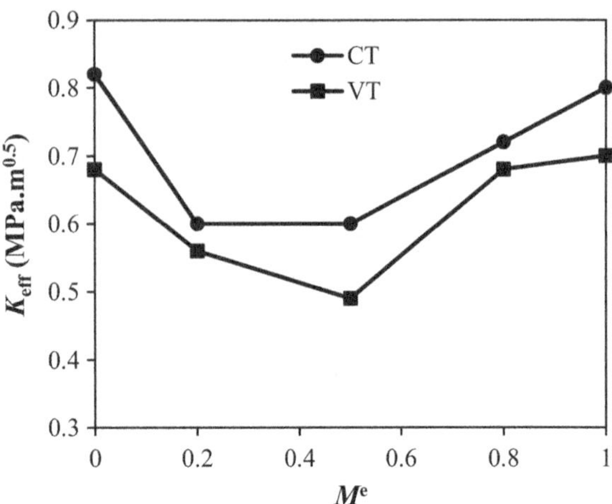

Figure 4.1 displays the fracture resistance of WMA mixtures at different environmental and loading conditions. According to this figure, the WMA concretes conditioned at the VT condition performed worse than those conditioned at the CT condition in terms of resistance against the crack growth under any mode of loading. The negative effect of VT condition on the fracture resistance may be attributed to the differential thermal contraction (DTC) damage occurred in WMA concrete. This damage has been described in Chap. 3 (Sect. 3.7). In addition, the minimum value of fracture resistance was achieved under mixed mode of $M^e = 0.5$, signifying that WMA concrete was exposed to a critical condition of fracture when it was subjected to a mixed mode I/II loading.

Razmi and Mirsayar [8] studied the fracture behavior of WMA concretes under different modes of loading and at three different temperatures. Limestone aggregates with gradation No. 4 (shown in Table 3.2) and binder with penetration grade of 60/70 (as a base binder) were used in this research. In order to prepare the WMA concretes, 3% Sasobit (by weight of the base binder) was blended with the base binder at 130 °C for 15 min using a high shear mixer operating at 4000 rpm.

They used the SCB specimen shown in Fig. 3.4a for performing the fracture tests. This specimen had a diameter of 150 mm, thickness of 32 mm, and crack length of 20 mm. Meanwhile, the locations of crack and bottom supports for performing the fracture tests under different modes of loading are given in Table 3.1.

In this study, the fracture tests were carried out under pure mode I, pure mode II, and three different mixed modes of $M^e = 0.8$, $M^e = 0.5$, and $M^e = 0.2$ at a displacement rate of 3 mm/min. In addition, the tests were conducted at three different temperatures of 0, -10, and -20 °C, such that the SCB specimens were put into a freezer adjusted at a certain temperature for 12 h, and were then loaded using a universal testing machine equipped with a three-point bend fixture (see Fig. 3.8).

Figure 4.2 displays the fracture resistance of WMA concretes at different temperature and loading conditions. Based on this figure, the fracture resistance of WMA

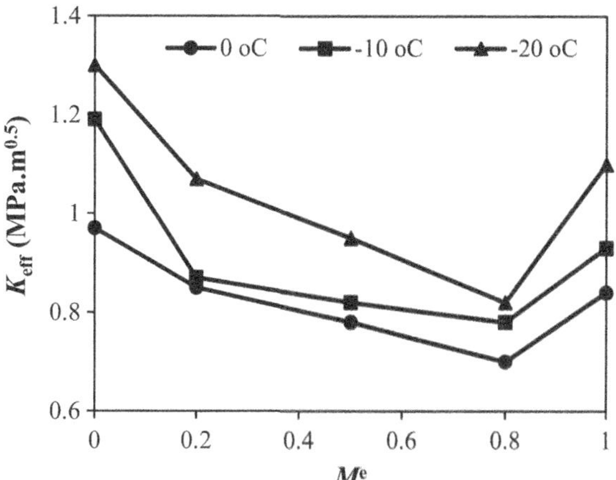

Fig. 4.2 Fracture resistance of WMA concrete at different temperature and loading conditions [8]

concrete (for all the modes of loading) increased as the temperature dropped. The similar behavior has been reported by Mansourian et al. [9] and Aliha et al. [12]. They investigated the fracture resistance of different WMA concretes (i.e., plain WMA and jute and FORTA fibers reinforced WMA) under different modes of loading (i.e., pure mode I, pure mode II, and two mixed modes of $M^e = 0.5$ and $M^e = 0.2$) at three different temperatures of 0, -10 and $-20\,°C$. According to the results of these investigations, the fracture resistance of all the investigated WMA concretes increased as the temperature reduced. This behavior was observed for all the modes of loading investigated in this research.

The increase of fracture resistance can be attributed to the fact that the binder stiffness increases as temperature decreases. It is worth noting that HMA concretes show the similar behavior, as discussed in Chap. 3. Furthermore, the minimum value of fracture resistance was achieved under mixed mode of $M^e = 0.8$.

It is worth mentioning that the crack growth paths in the SCB specimens tested under different modes of loading were similar to those observed for the HMA mixture shown in Fig. 3.18, in which for the pure mode I loading, the crack initiated along the initial crack line, and developed straightly toward the top fixture whilst for the mixed mode I/II and pure mode II loadings, the crack kinked from the initial crack line, and propagated along a curvilinear path. This manner of the crack growth path can be attributed to the maximum tensile stress around the crack front, which was not anymore along the initial crack as the SCB specimen was loaded under mixed mode I/II or pure mode II [13]. Meanwhile, as the proportion of shear mode (i.e., mode II) at the crack front of the SCB specimen increased (i.e., as M^e decreased), the initiation angle of fracture became greater. It is also pointed out that the crack propagated through the aggregates indicating the brittle fracture behavior of WMA concretes at low temperatures. Whereas, the crack propagates around aggregates at the interface between aggregates and binder as the tests are performed at relatively high temperatures [14].

In other investigations, Lee et al. [15] and Yeon et al. [16] studied the fracture resistance of various WMA concretes at different low temperatures (i.e., -5, -15, -20, -25, and $-35\,°C$). Three WMA additives (i.e., Evotherm, Pewo, and Sasobit) were used for preparing different WMA concretes. In this research, SENB tests were performed using a three-point bend fixture to measure the fracture toughness (K_{If}) at -5, -15, -20, -25, and $-35\,°C$. For all the WMA mixtures, the fracture toughness initially increased to reach its peak value at $-15\,°C$ or $-20\,°C$, and then decreased as the test temperature reduced. The decrease in the fracture toughness is attributed to the DTC damage, as mentioned earlier. It is noticed that Podolsky et al. [17] also investigated the fracture toughness and fracture energy of WMA concretes containing different additives using the SCB tests at three low temperatures of 0, -12, and $-24\,°C$. According to their results, the fracture energy increased as the temperature enhanced. In addition, the fracture toughness increased by reduction of the temperature from 0 to $-12\,°C$, and further reduction in the temperature from -12 to $-24\,°C$, decreased the fracture toughness due to the DTC damage.

4.2.2   Effect of Crumb Rubber

Razmi and Mirsayar [8] studied the effect of crumb rubber on the fracture resistance of WMA concrete at different temperature and loading conditions. In order to prepare modified WMA concrete, 20% of the crumb rubber particles were blended with the base binder containing 3% Sasobit at 190 °C for 30 min using a high shear mixer operating at 2000 rpm followed by 45 extra minutes with 1000 rpm. All other specifications of the base binder, aggregates, test specimen, fracture tests, etc, were given in Sect. 4.2.1.

Figure 4.3 exhibits the fracture resistance of the WMA concrete modified by crumb rubber particles at different temperature and loading conditions. Comparing

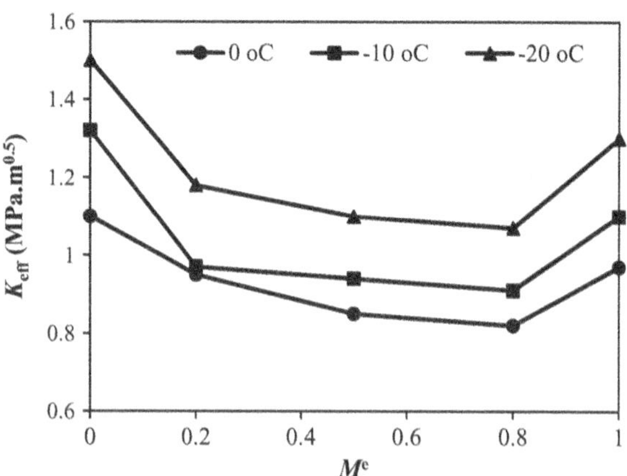

Fig. 4.3 Fracture resistance of the modified WMA concrete at different temperature and loading conditions [8]

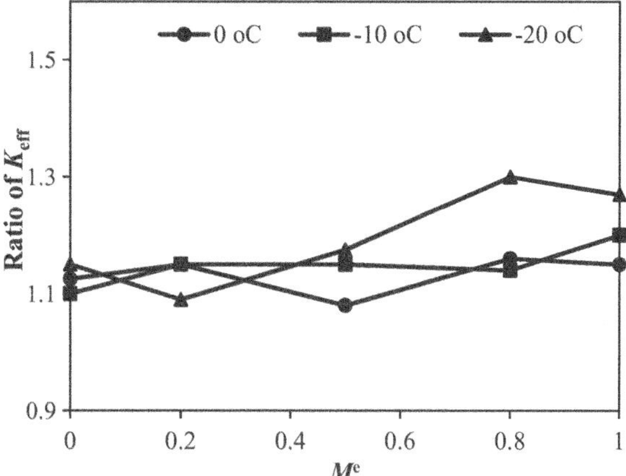

Fig. 4.4 Ratio of K_{eff} for the crumb rubber modified WMA to that for the plain WMA [8]

the results of the modified WMA concrete (shown in Fig. 4.3) and plain WMA concrete (shown in Fig. 4.2) indicates that the crumb rubber improved the fracture resistance of WMA concrete considerably.

The ratio of K_{eff} for the crumb rubber modified WMA to that for the plain WMA at different temperature and loading conditions is shown in Fig. 4.4. According to this figure, the positive effect of crumb rubber was more considerable at the lowest temperature. Furthermore, the same improvement in fracture resistance for all the temperature and loading conditions was nearly observed by addition of the crumb rubber to WMA concrete, except for the case of $-20\,°C$.

In another study, Pirmohammad and Khanpour [10] investigated the effect of crumb rubber on the fracture resistance of WMA concretes. The WMA mixtures considered in this research contained aggregates with gradation No. 4, binder with penetration grade of 60/70 (as a base binder), and air void of 4%. Meanwhile, the modified WMA concrete contained 15% crumb rubber (by weight of the base binder).

In order to prepare the crumb rubber modified binder, particles of the crumb rubber were added to the base binder at 170 °C, and were then mixed for 240 min using a high shear mixer operating at 5500 rpm.

Figure 4.5 shows the fracture resistance of WMA concrete modified by crumb rubber at different environmental and loading conditions. Similar to the results of plain WMA concrete shown in Fig. 4.1, the fracture resistance of the modified WMA concrete conditioned at the VT was also less than that of the modified WMA concrete at the CT under any mode of loading. Furthermore, comparing the results of the modified WMA concrete (shown in Fig. 4.5) and plain WMA concrete (shown in Fig. 4.1) indicates that the crumb rubber improved the fracture resistance of WMA concrete significantly, such that the fracture resistance increased 19 and 17% at the CT and VT conditions, respectively. Indeed, the crumb rubber imparts ductility to the base binder, and also it improves the adhesion between binder and aggregates [18]. It is also pointed out that the minimum value of fracture resistance was achieved under mixed mode of $M^e = 0.5$.

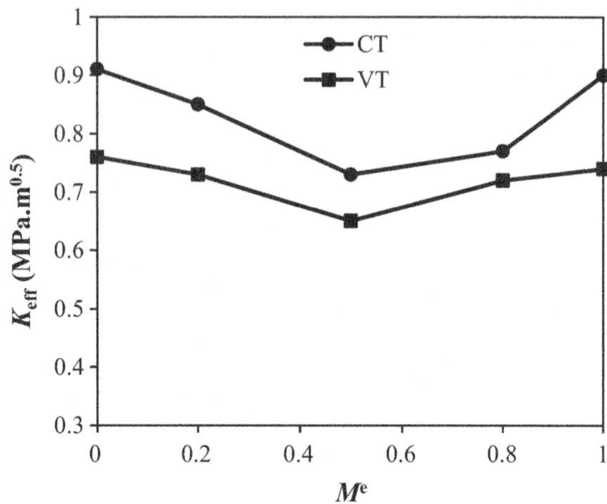

Fig. 4.5 Fracture resistance of the modified WMA concrete at different environmental and loading conditions [10]

4.2.3 Effect of Fiber

Mansourian et al. [9] investigated the effect of jute fiber on the fracture resistance of WMA concrete under different temperature and loading conditions. Binder with penetration grade of 80/100 was used in this research as a base binder, and aggregates with gradation No. 4 were employed. In addition, all WMA mixtures considered in this research were prepared with optimum binder content and air void content of 5.6% and 4%, respectively.

In order to prepare the plain WMA concrete, 3% Sasobit (by weight of the base binder) was blended with the base binder at 130 °C for 10 min using a low shear mixer working at 300 rpm.

As such, jute fibers with 20 mm length at three different dosages of 0.3, 0.5, and 0.7% (by weight of total WMA concrete) were used for preparing jute fiber reinforced WMA concretes. As discussed in Chap. 3 (Sect. 3.9), there are three methods of wet, dry, and combination of them to incorporate fibers into asphalt mixtures. In this research, the dry method was used for preparation of the reinforced WMA concretes, such that the jute fibers with a certain dosage were initially blended with the hot aggregates, and the blend was then mixed with the binder (i.e., base binder + 0.3% Sasobit) by laboratory mixer at 130 °C.

They used the SCB specimen shown in Fig. 3.4a for conducting fracture experiments under pure mode I, pure mode II, and two different mixed modes of $M^e = 0.5$ and $M^e = 0.2$. The SCB specimen had a diameter of 150 mm, thickness of 32 mm, and crack length of 20 mm. In addition, the tests were conducted at three different temperatures of 0, −10, and −20 °C, such that the SCB specimens were put into a freezer adjusted at a certain temperature for 12 h, and were then loaded using a universal testing machine equipped with a three-point bend fixture (see Fig. 3.8) at a displacement rate of 3 mm/min. Meanwhile, the locations of crack and bottom supports for performing the fracture tests under different modes of loading are given in Table 3.1.

Figure 4.6 exhibits the fracture resistance (K_{eff}) of the plain and jute fiber reinforced WMA concretes at different temperature and loading conditions. Based on the results, the addition of jute fibers to mixture increased the fracture resistance of WMA concrete under pure mode I and mixed mode of $M^{\text{e}} = 0.5$. Application of 0.3% jute fiber improved the mode I fracture resistance by 35%. The positive effect

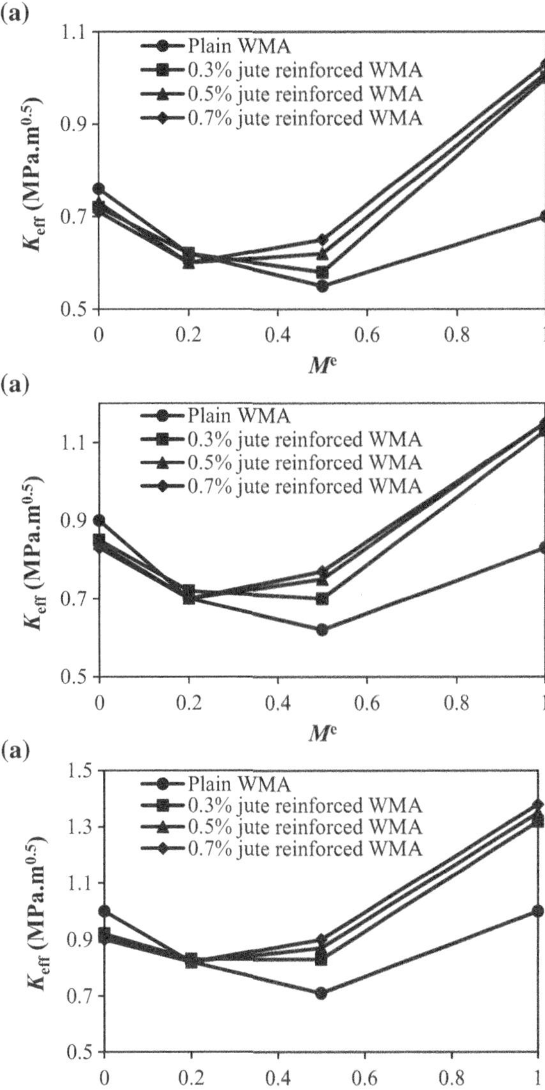

Fig. 4.6 Fracture resistance of WMA concretes under different modes of loading at **a** 0 °C, **b** −10 °C, −20 °C [9]

Table 4.2 Physical and mechanical properties of fibers [9]

Fiber type	Specific gravity (g/cm^3)	Tensile strength (MPa)	Length (mm)	Color	Melting point (°C)
Jute	–	200–450	20	Gray	–
FORTA	0.91–1.44	490–2800	20	Black-Yellow	100–427

of jute fibers on mode I (i.e., tensile mode) fracture resistance may be attributed to the improved tensile strength of WMA mixtures due to increasing the binder viscosity and forming good adhesion between the jute fibers and binder. Indeed, fibers absorb the light components of binder and increase the binder viscosity [9, 19]. Previous researches have shown that increase in the binder viscosity enhances the tensile strength of asphalt mixture [9, 20].

Another result from Fig. 4.6 is that the jute fibers did not improve the fracture resistance of WMA concrete under dominant mode II loadings (i.e., mixed mode of $M^e = 0.2$ and pure mode II), and they had even negative effect on the pure mode II fracture resistance. Meanwhile, the addition of jute fibers more than 0.3% did not have a significant effect on the fracture resistance, so the optimum content of jute fibers seems to be around 0.3%.

In another investigation, Aliha et al. [12] studied the effect of jute (as a natural fiber) and polyolefin–aramid (as a synthetic fiber) fibers on the fracture behavior of WMA concrete at different temperature and loading conditions. Binder with performance grade (PG) of 52–28 and penetration grade of 80/100 was used as a base binder. In addition, limestone aggregates with gradation No. 4 and air void content of 4% were employed in the preparation of mixtures.

Both the jute and polyolefin–aramid (FORTA) fibers with 20 mm lengths at three different dosages of 0.3, 0.5, and 0.7% (by weight of total WMA concrete) were incorporated into WMA concrete to prepare fiber reinforced WMA concretes. Similar to Mansourian et al. [9], the dry procedure was employed in this research to prepare the fiber reinforced WMA concretes, as explained above. Table 4.2 presents the physical and mechanical properties of the fibers used in this research. It is pointed out that the polyolefin–aramid fiber used in this research was produced by FORTA Company in USA.

The SCB specimen used by Mansourian et al. [9] was employed in this research [12] for performing the fracture tests under pure mode I, pure mode II, and two different mixed modes of $M^e = 0.5$ and $M^e = 0.2$. In addition, the tests were conducted at three different temperatures of 0, −10, and −20 °C. It is useful to note that the test procedure adopted by Aliha et al. [12] was similar to the one employed by Mansourian et al. [9].

Figure 4.7 shows the effect of jute and FORTA fibers on the fracture resistance of WMA concrete at different temperatures. According to this figure, FORTA fibers improved the fracture resistance of WMA concrete better than jute fibers; however, increase in the fracture resistance was more pronounced as the portion of opening mode was dominant at the crack front (i.e., mixed mode of $M^e = 0.5$ and pure mode I). For the cases of dominant shear modes (i.e., mixed mode of $M^e = 0.2$ and pure

Fig. 4.7 Fracture resistance of WMA concretes under different modes of loading at **a** 0 °C, **b** −10 °C, −20 °C [12]

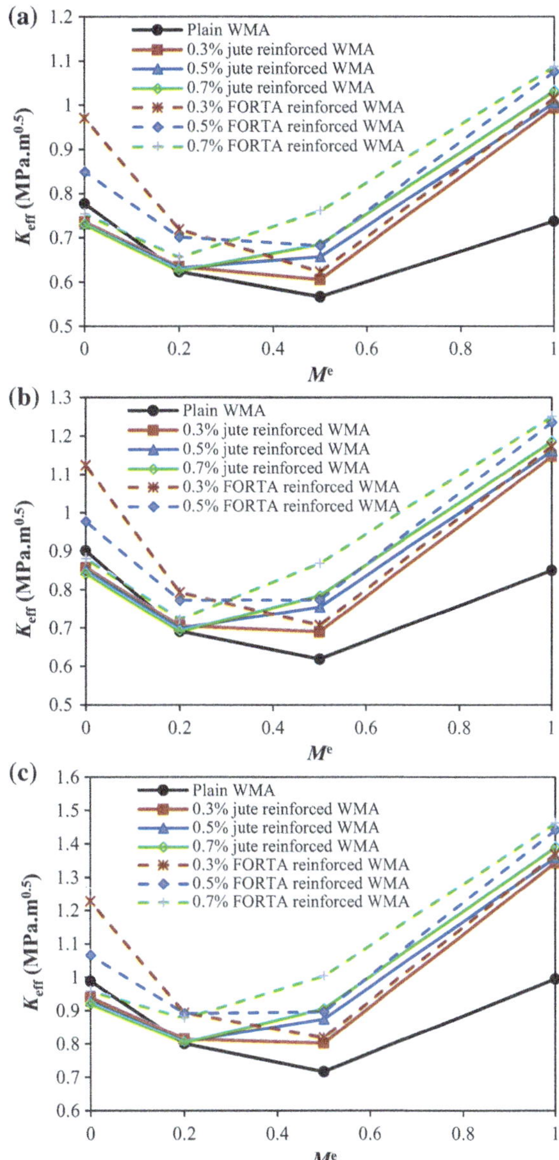

mode II), the jute fibers with any dosage and FORTA fibers with a dosage of 0.7% did not have a sound effect on the fracture toughness. Meanwhile, the trends of fiber effect on the fracture toughness were similar for all the tests temperatures.

An interesting result from Fig. 4.7 is that as the dosage of FORTA fibers enhanced, the fracture toughness measured under the dominant mode I loading increased whilst the fracture toughness measured under the dominant mode II loading decreased.

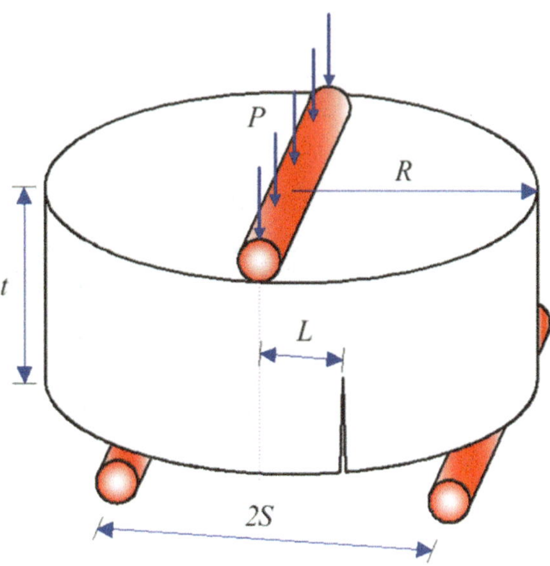

Fig. 4.8 Disk bend specimen [24]

The similar results have been reported by Fazaeli et al. [21] when investigating the effect of polyolefin–aramid (FORTA) fiber on the mode I fracture resistance of WMA concretes. According to their results shown in Fig. 4.8, the fiber reinforced mixtures demonstrated higher fracture toughness than plain mixtures. Since the melting point of Polyolefin is about 100 °C [21], the Polyolefin available in FORTA fibers melted at the temperature of preparing asphalt mixture and acted as a modifier of binder [22]. Furthermore, aramid fibers due to having high tensile strength (i.e., 2758 MPa) were responsible to enhance the fracture resistance of asphalt mixture. Details of this investigation would be discussed in the subsequent section.

Figure 4.7 also shows the effect of loading mode on the fracture resistance of the WMA concretes, in which the fracture toughness decreased and then increased as the shear mode enhanced.

Pirmohammad and Hojjati [23] investigated the effects of kenaf and basalt fibers on fracture toughness of WMA concrete under different modes of loading. Disk bend specimen [24] shown in Fig. 4.8 was used in this study. This specimen had a radius of 50 mm, thickness of $t = 40$ mm, and crack length of $a = 16$ mm. All modes of loading ranging from pure mode I to pure mode II can be simulated using this specimen by simply translating the crack position (i.e., L). Pure mode I and pure mode II were achieved as the crack was located at $L = 0$ and $L = 38$ mm, respectively. Other than pure modes I and II, the fracture tests were also performed at two mixed modes of I/II (i.e., $M^e = 0.79$ and $M^e = 0.35$) in this research. The crack was located at $L = 29$ mm and $L = 36$ mm for the mentioned mixed modes of $M^e = 0.79$ and $M^e = 0.35$, respectively. Figure 4.9 displays the values of geometry factors (i.e., Y_I and Y_{II}) obtained from finite element analyses.

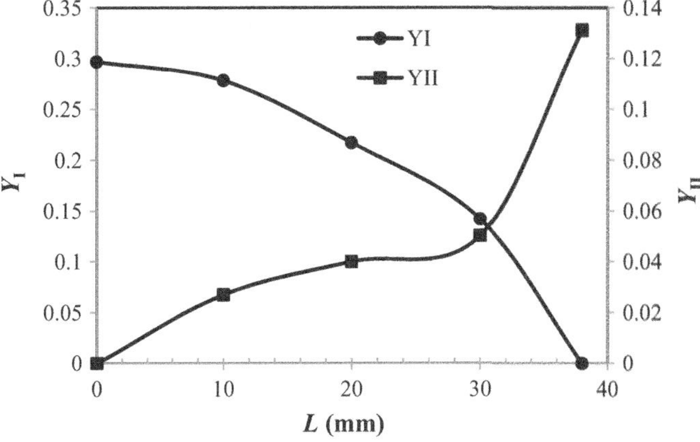

Fig. 4.9 Values of geometry factors for the disk bend specimen [24]

The WMA concretes studied in this research contained a base binder with 60/70 penetration grade and 3% Sasobit. In addition, limestone aggregates with gradation No. 5 were used in the preparation of WMA concretes. Air void content of the WMA concretes was about 4%.

Kenaf and basalt fibers with a length of 0.8 mm and different dosages of 0.1, 0.3, and 0.5% (by weight of the total WMA mixture) were incorporated into WMA mixture. Table 4.3 presents the properties of kenaf and basalt fibers used in this research. It is pointed out that another WMA concrete (i.e., control WMA) was also prepared to highlight the effect of fibers on the fracture behavior of mixtures.

The disk bend specimens made of the abovementioned mixtures were kept in a freezer set at $-15\,°C$, and fracture tests were then performed using a three-point bend fixture with a span of $2S = 80$ mm and displacement rate of 3 mm/min. Fracture load P_{cr} measured from the tests and the geometry factors shown in Fig. 4.9 are given in the following equations to calculate the fracture resistance of the WMA concretes (i.e., K_{eff}).

$$K_{If} = \frac{6P_{cr}S}{Rt^2}Y_I\sqrt{\pi a} \tag{4.1}$$

$$K_{IIf} = \frac{6P_{cr}S}{Rt^2}Y_{II}\sqrt{\pi a} \tag{4.2}$$

Table 4.3 Properties of kenaf and basalt fibers [11]

Fiber type	Elastic modulus (GPa)	Tensile strength (MPa)	Diameter (μm)	Density (g/cm^3)
Kenaf	53	930	50–60	1.45
Basalt	85	2800	10–13	2.67

$$K_{\text{eff}} = \sqrt{K_{\text{If}}^2 + K_{\text{IIf}}^2} \qquad (4.3)$$

It is reminded that the value of K_{eff} reduces to K_{If} and K_{IIf} for the cases of pure mode I and pure mode II, respectively.

Figure 4.10 shows the fracture resistance of the control and fiber reinforced WMA mixtures. Based on the results, both the kenaf and basalt fibers increased the fracture resistance of WMA concrete under any mode of loading. However, the kenaf fibers performed better than the basalt fibers as the specimens were subjected to dominant mode I loading (i.e., pure mode I and mixed mode of $M^e = 0.79$), while this result reversed for the case of dominant mode II loading i.e., the basalt fibers showed higher fracture resistance than kenaf fibers as the specimens were loaded under dominant mode II (i.e., mixed mode of $M^e = 0.35$ and pure mode II). Furthermore, the addition of 0.3% kenaf and basalt fibers demonstrated the best results as compared with other dosages (i.e., 0.1 and 0.5%) of kenaf and basalt fibers considered in this research.

4.2.4   Effect of Aggregate Type

Fazaeli et al. [21] studied the effect of aggregate type on the fracture toughness of WMA concretes. For this purpose, laboratory and field samples were prepared in this research. The laboratory samples were made of siliceous aggregates (i.e., river materials); whereas, the filed samples were made of limestone aggregates (i.e., mountain materials). The laboratory samples were prepared from three different asphalt concretes including unmodified asphalt concrete, asphalt concrete modified by 2% Sasobit (by weight of the base binder), and asphalt concrete modified by both Sasobit (2% by weight of the base binder) and FORTA fibers (0.5 kg per ton of asphalt mixture). Furthermore, the field samples were prepared from two asphalt concretes including unmodified asphalt concrete and asphalt concretes modified by Sasobit (1 kg per ton of asphalt mixture) and FORTA fibers (0.5 kg per ton of asphalt mixture). For convenience, the mentioned five asphalt concretes investigated in this research were designated as S unmodified, SW, SFW, L unmodified, and LFW, respectively. The letters S and L correspond to the siliceous and limestone aggregates, and F and W refer to the fiber and warm mix asphalt concretes.

Both the siliceous and limestone aggregates were used in accordance with gradation No. 4 (presented in Table 3.2) in the preparation of samples. In addition, two types of binder were utilized for providing samples: (i) pure binder with a penetration grade of 66 and performance grade (PG) of 58-22 was used in the unmodified samples, (ii) modified binder (containing pure binder + 2% Sasobit) with a penetration grade of 47 and PG of 70-16 was used for providing modified asphalt concretes. It is worth noting that the addition of 2% Sasobit to the binder decreased the penetration grade, and, therefore, caused the PG 58-22 binder to behave similar to the PG 70-16 binder.

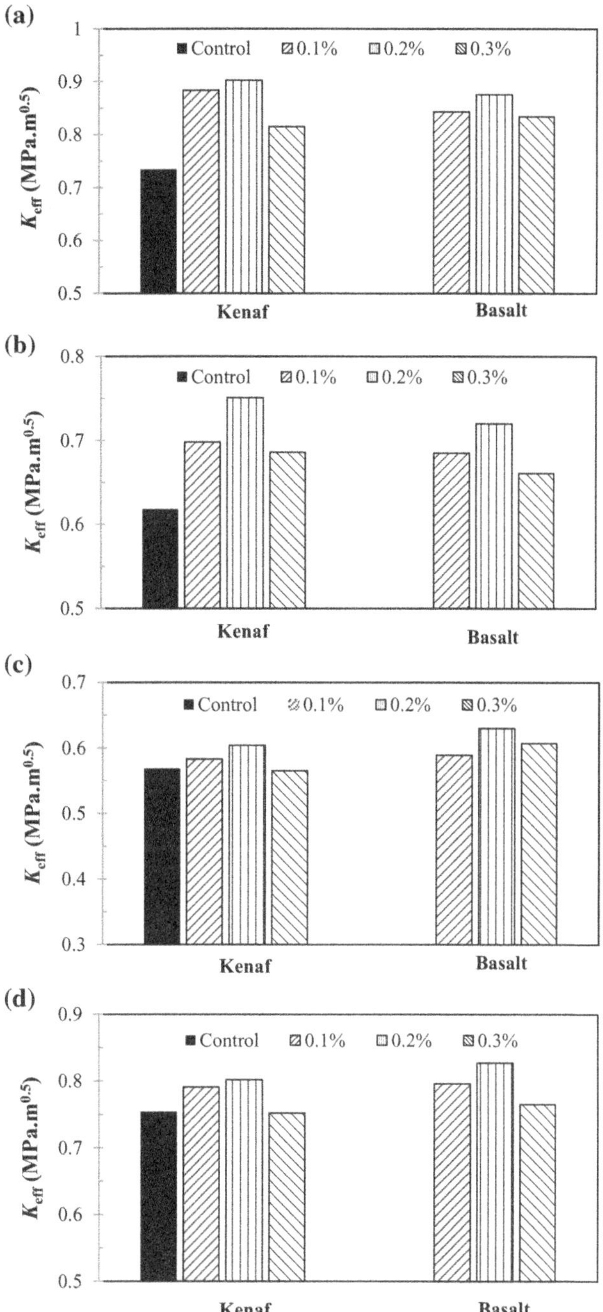

Fig. 4.10 Fracture resistance of control and fiber reinforced WMA concretes under **a** pure mode I, **b** mixed mode of $M^e = 0.79$, **c** mixed mode of $M^e = 0.35$, and **d** pure mode II [23]

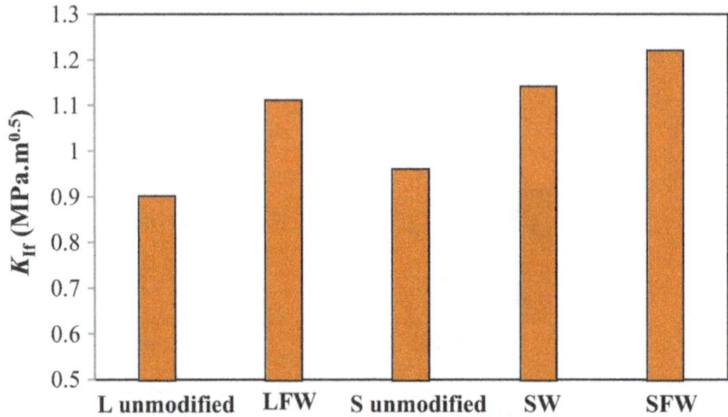

Fig. 4.11 Fracture toughness of asphalt mixtures [21]

The same amount of binder was used for preparing the unmodified and modified asphalt mixtures. These amounts were 5.4 and 5% for asphalt mixtures containing siliceous and limestone aggregates, respectively. The Sasobit modified binder is prepared by mixing Sasobit and binder at 130 °C for 5 min using a mixer operating at 300 rpm.

The FORTA fibers with 19 mm in length were used in this research. Furthermore, the dry method was used for incorporating the fibers into asphalt mixture. It is pointed out that by addition of Sasobit, the aggregate and binder temperatures were reduced from 170 and 145 °C to 150 and 130 °C, respectively. Meanwhile, both the laboratory and field samples contained an air void of 4–5% depending on the type of asphalt mixture.

The SCB specimen with a diameter of 150 mm, thickness of 32 mm, and crack length of 20 mm was used to conduct the fracture tests under pure mode I loading. In order to perform the fracture tests, the SCB specimens were put into a freezer adjusted at 0 °C for 6 h, and were then loaded using a universal testing machine equipped with a three-point bend fixture at a displacement rate of 3 mm/min. Meanwhile, the bottom supports were set at $S_1 = S_2 = 50$ mm during the mode I experiments.

Figure 4.11 displays the fracture toughness of asphalt mixtures containing different aggregate types. According to the results, all the modified and unmodified mixtures containing siliceous aggregates demonstrated higher fracture toughness than those containing limestone aggregates.

4.3   Comparison of Fracture Resistance of WMA and HMA Concretes

Hill et al. [1] compared fracture energy of different WMA mixtures with that of HMA mixture using DC(T) tests at −12 °C. Binder with performance grade (PG) of 64-22 (as a base binder) was used in this research to produce WMA and HMA concretes.

The WMA technologies investigated in this research included at least one additive from each of the three WMA categories stated in Sect. 4.1. Four different WMA concretes were prepared using the additives of Sasobit, Evotherm M1, Advera, and Rediset LQ.

Sasobit is a synthetic paraffin wax. This additive stiffens binder and enhances the high-temperature performance grade. For example, the addition of 2.5% Sasobit to a binder with PG 58-28 causes it to behave similar to a binder with PG 64-22 [25]. Advera is a synthetic zeolite containing nearly 20% entrapped water by weight in its microstructure. This water is released at temperatures above 100 °C, leading to foaming of the binder. Indeed, the considerable volume expansion resulting from the foaming process together with the low viscosity of the foamed binder leads to produce the mixture at lower temperatures. Evotherm M1 is a type of the Evotherm 3G liquid chemical additive manufactured by Meadwestvaco. This additive does not contain water. Rediset LQ is the liquid form of the chemical additive manufactured by Akzo Nobel. Both Rediset LQ and Evotherm operate as surfactants and are potentially able to better the stress relaxation properties of asphalt mixture. It is notable that Sasobit and Advera correspond to the organic and foaming categories of WMA additives, respectively, and both the Evotherm M1 and Rediset LQ refer to chemical category of WMA additives.

The optimum binder and air void contents were considered 6.7% and 4%, respectively, for all the WMA and HMA concretes. The mixing and compaction temperatures for the HMA concrete were 160 °C and 150 °C, respectively, whilst those for the WMA concretes were 135 and 125 °C.

The DC(T) tests were performed on the mixtures as per ASTM 7313-07. The fracture experiments were conducted at −12 °C, as recommended in ASTM for mixtures containing PG 64-22 binder (i.e., 10 °C warmer than the low-temperature grade of the PG). The DC(T) tests were also performed at a CMOD rate of 1 mm/min. More information on the DC(T) test and the procedure of calculating fracture energy can be found in Chap. 5. It is also pointed out that four replicates were tested for each mixture.

Hill et al. [1] obtained the fracture energy of HMA, WMA-Sasobit, WMA-Adver, WMA-Evotherm, and WMA-Rediset LQ as 377, 329, 335, 404, and 404 J/m^2, respectively, in which the HMA concrete demonstrated higher fracture energy than the WMA concretes containing Sasobit and Advera additives, while the fracture energy of the HMA concrete was smaller than that of the WMA concretes containing chemical additives including Evotherm and Rediset LQ. Thus, the chemical additives improved the fracture energy of asphalt mixture as compared with the HMA, whereas the organic and foaming additives adversely influenced the fracture resistance of asphalt mixture.

In another study, Yoo et al. [26] compared the pure mode I fracture toughness of HMA and WMA concretes using SENB test. Two WMA additives including the liquid form of Evotherm (EV) and solid form of paraffin wax (PW) were employed in this research to produce WMA mixtures. The contents of EV and PW were 0.5 wt% and 1 to 2 wt% of the total binder, respectively.

Binder with penetration grade of 60/80 and PG of 64-22 (as a base binder) and air void of about 4% were used in this research to produce WMA and HMA concretes.

Different polymers including low-density polyethylene (LD), ethylene-vinyl acetate (VA), and ethylene propylene diene monomer (EPDM) rubber (or diene monomer (DM)) were used to prepare asphalt mixtures. Table 4.4 presents the mixtures considered in this research with different compositions. Meanwhile, the contents of each additive are also given in Table 4.4.

EV and PW were melted into the binder heated at 160 °C and blended using a spatula for 3–5 min before use. Powders of LD and VA were slowly poured into binder heated at 170 °C and blended using a high shear mixer operating at 4000 rpm for 60 min before use, and at this point, DM was added, if needed.

The SENB specimen shown in Fig. 4.12 was prepared and put into an environmental chamber set at −20 °C for 12 h, and was then tested using a three-point bend fixture with a displacement rate of 3.3 mm/min.

Yoo et al. [26] obtained the fracture toughness of HMA-CON, HMA-LV, and HMA-LVD as 1.23, 1.36, and 1.34 MPa.m$^{0.5}$. Likewise, that of WMA-PW 1.0, WMA-EV 0.5, WMA-PW 1.0 LV, WMA-EV 0.5 LV, WMA-PW 2.0 LVD and WMA-EV 0.5 LVD was measured as 1.24, 1.32, 1.3, 1.38, 1.37 and 1.47 MPa.m$^{0.5}$. According to the results, fracture toughness of the unmodified WMA concrete (i.e., PW 1.0) was found to be somewhat higher than that of the unmodified HMA concrete (i.e., CON). Generally, polymer modified HMA and WMA concretes showed higher resistance against fracture than unmodified ones. The addition of DM to the HMA mixture containing LD and VD (i.e., the HMA containing LVD) adversely affected the fracture toughness. In contrary to the HMA concrete, the addition of DM to the

Table 4.4 Contents of asphalt mixtures [26]

Mixture type	Designation	Additive (%)		Polymer (%)			PG
		PW	EV	LD	VA	DM[a]	
HMA							
	CON	0	0	0	0	0	64–22
	LV	0	0	2	3	0	76–22
	LVD	0	0	2	3	10	76–22
WMA							
	PW 1.0	1	0	0	0	0	64–22
	EV 0.5	0	0.5	0	0	0	64–22
	PW 1.0 LV	1	0	2	3	0	70–22
	EV 0.5 LV	0	0.5	2	3	0	70–22
	PW 2.0 LVD	2	0	2	3	10	76–22
	EV 0.5 LVD	0	0.5	2	3	10	76–22

[a]Percent by weight of LD + VA.

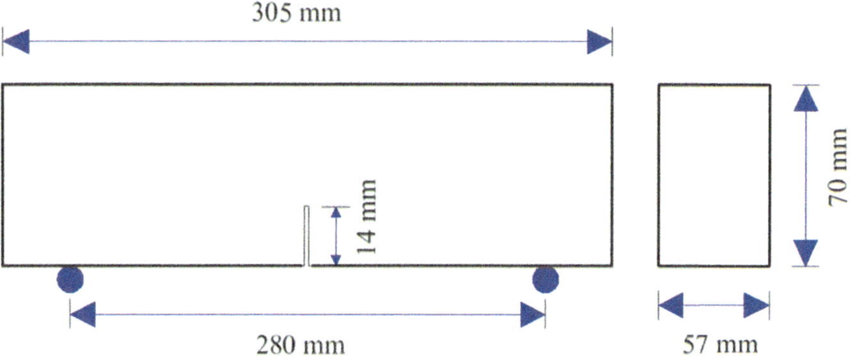

Fig. 4.12 Geometry and dimensions of SENB specimen [26]

WMA concrete improved the fracture toughness. For the case of WMA concretes, the use of EV performed better than PW. Considering the results given above reveals that the fracture toughness of WMA was similar to, or somewhat higher than, that of HMA. Furthermore, the WMA concrete designated as EV 0.5 LVD demonstrated the highest fracture toughness as compared with other WMA and HMA concretes investigated in this study.

Lee et al. [15] and Yeon et al. [16] have also compared the fracture toughness of various WMA and HMA concretes at different low temperatures (i.e., $-5, -15, -20, -25,$ and $-35\,^{\circ}C$). Three WMA additives (i.e., Evotherm, Pewo, and Sasobit) and two polymers (LVM and SBS) were used for preparing different WMA and HMA concretes. It is reminded that the LVM is a compound of low-density polyethylene (LDPE), ethylene-vinyl acetate (EVA), and ethylene propylene dine monomer (EPDM). The SENB tests were performed using a three-point bend fixture to measure the mode I fracture toughness (K_{If}) at $-5, -15, -20, -25,$ and $-35\,^{\circ}C$. Most of the WMA concretes studied in this research demonstrated higher fracture toughness than HMA at severely low temperatures (i.e., -25 and $-35\,^{\circ}C$). As a result, WMA mixtures were found to be less susceptible to brittle fracture than HMA mixtures.

In an investigation performed by Singh et al. [27], the fracture resistance of HMA and WMA mixtures containing different dosages of reclaimed asphalt pavement (RAP), i.e., 0, 10, 20, 30, and 40%, have been studied. Five different HMA (including HMA-0%RAP, HMA-10%RAP, HMA-20%RAP, HMA-30%RAP, and HMA-40%RAP) and WMA (including WMA-0%RAP, WMA-10%RAP, WMA-20%RAP, WMA-30%RAP, and WMA-40%RAP) mixtures were prepared using basaltic aggregates with 19 mm NMAS. In addition, the target air void was 6.5% in the mixtures preparation.

Binder with penetration grade of 60/70 (as a base binder) and Fischer–Tropsch (FT) wax (as a WMA additive) were used in this research. The content of FT wax was 2% by weight of the base binder for producing WMA mixtures.

The SCB specimens with 150 mm in diameter and 50 mm in thickness were prepared by Marshal compactor machine. Notch with three different depths of 20, 25, and 32 mm, and width of 2 mm were generated at the middle of the SCB specimens.

Fracture tests were performed on the SCB specimens using a three-point bend fixture at 35 °C with a displacement rate of 0.5 mm/min. The strain energies were calculated from load-LLD curves recorded from the tests. In other words, the strain energy for each notch depth was obtained by computing the area under the load–displacement curve to the failure point. Critical J-integral (i.e., J_c) was finally obtained using Eq. (5.6). It is pointed out that the procedure of calculating critical J-integral has been discussed in Chap. 5.

Figure 4.13 plots the values of J_c calculated for the HMA and WMA concretes containing different dosages of RAP. On the basis of the results, both the plain HMA and WMA concretes (i.e., mixtures without RAP) showed similar resistance to fracture. However, HMA concretes were found more fracture resistant compared to WMA concretes containing the same dosages of RAP. This may be attributed to the stiffer nature of WMA mixtures as compared with HMA ones at intermediate temperatures. Moreover, the addition of RAP to HMA and WMA mixtures increased the value of J_c.

Pirmohammad et al. [28, 29] investigated fracture resistance of WMA concretes at different temperature and loading conditions using SCB tests. The SCB specimen had a diameter of 150 mm, thickness of 30 mm, and crack length of 20 mm.

Three different mixtures including HMA, WMA, and WMA-0.25%RAP were prepared in this research. Binder with 60/70 penetration grade was used as a base binder in this research to prepare these mixtures, and 0.3% Sasobit was added to the base binder to prepare WMA mixtures. In addition, 25%RAP was used in the preparation of the mixture designated as WMA-0.25%RAP. It is worth noting that superpave mix design was used to prepare the mixtures. Siliceous aggregates with

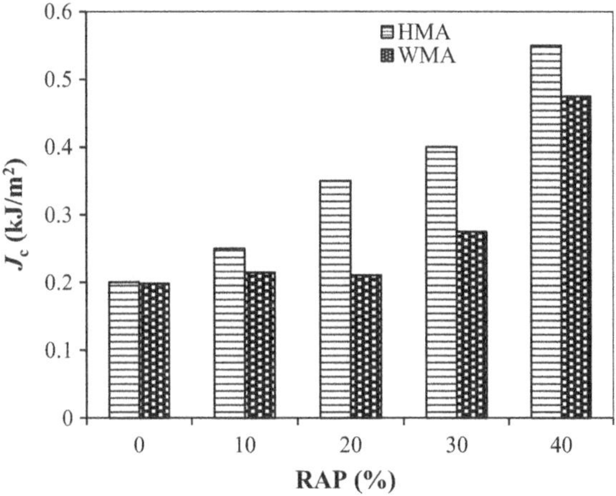

Fig. 4.13 Values of J_c for HMA and WMA concretes containing different dosages of RAP [27]

gradation shown in Fig. 4.14 and air void content of 4% were employed for the preparation of mixtures.

Fracture experiments were performed under pure mode I, pure mode II, and a mixed mode of $M^{e} = 0.5$. The crack location (L) and bottom support locations (S_1 and S_2) in the SCB tests for simulating different modes of loading together with the corresponding values of the geometry factors (Y_{I} and Y_{II}) are given in Table 4.1. The SCB tests were conducted at two temperatures of -15 and $25\,^{\circ}$C.

Figures 4.15 and 4.16 display the fracture energy and fracture toughness of various HMA and WMA concretes at different temperature and loading conditions. Based on the results given in Fig. 4.15, the WMA and WMA-25%RAP mixtures showed higher fracture toughness than the HMA mixture under any mode of loading, and

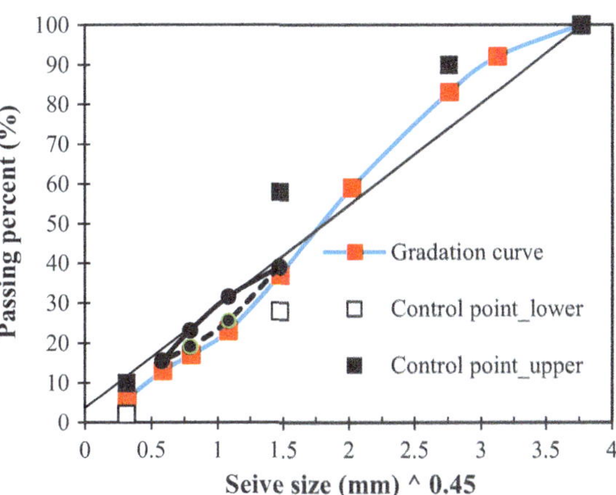

Fig. 4.14 Aggregate gradation curve of mixtures [29]

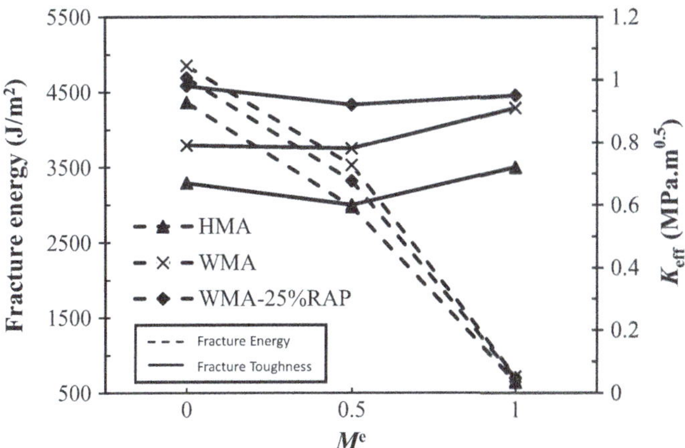

Fig. 4.15 Fracture energy and fracture toughness of asphalt mixtures under different modes of loading at $-15\,^{\circ}$C [29]

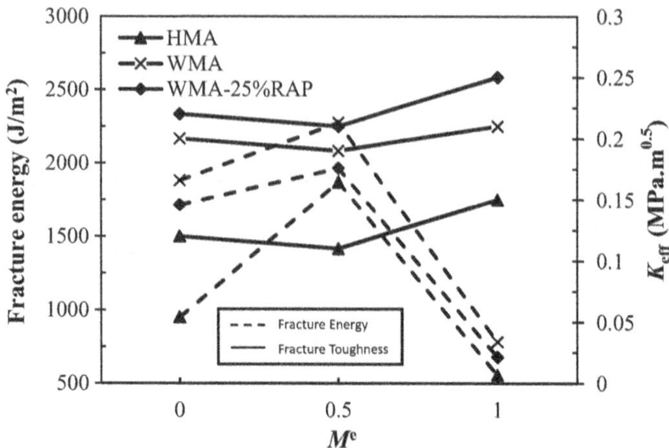

Fig. 4.16 Fracture energy and fracture toughness of asphalt mixtures under different modes of loading at 25 °C [29]

the WMA mixture containing 25%RAP (i.e., WMA-25%RAP) demonstrated the highest fracture toughness at −15 °C. Furthermore, the WMA concretes (i.e., WMA and WMA-25%RAP) resisted better than the HMA concrete to fracture under any mode of loading; however, the WMA concrete showed the highest value of fracture energy as compared with HMA and WMA-25%RAP, and the addition of RAP to WMA mixture resulted in the reduction of fracture energy. It is also pointed out that the mode I fracture energy of mixtures was nearly the same; while, the effect of Sasobit was more pronounced under shear loading (i.e., mixed mode of $M^e = 0.5$ and pure mode II).

According to Fig. 4.16, the trend of results of fracture toughness at 25 °C is similar to that observed for the test temperature of −15 °C. In other words, the WMA and WMA-25%RAP concretes exhibited higher fracture toughness than the HMA concrete, and the WMA concrete containing 25%RAP (WMA-25%RAP) showed the highest fracture toughness. Furthermore, the WMA concretes (i.e., WMA and WMA-25%RAP) performed better than the HMA concrete under any mode of loading; however, the WMA concrete showed the highest value of fracture energy as compared with the HMA and WMA-25%RAP concretes.

4.4   Summary

In this chapter, fracture behavior of WMA concretes was discussed. The effects of different parameters including mode of loading, temperature, crumb rubber, fiber, and aggregate type on fracture resistance of WMA mixtures were reviewed. The important results are listed below:

- Fracture resistance of WMA concretes decreases as the amount of shear load at crack front increases, and minimum resistance of WMA concretes to fracture is achieved under a mixed mode I/II loading not under a pure mode (i.e., I or II).
- As temperature drops, fracture resistance of WMA concretes increases due to the increase of binder stiffness, and further decrease in temperature results in reduction of fracture toughness due to the DTC damage. Furthermore, temperature variations lead to reduction in fracture resistance of WMA concrete.
- The use of crumb rubber has a positive effect on fracture toughness of WMA concrete.
- Different types of fiber such as jute, FORTA, kenaf, and basalt were incorporated to WMA mixture, and the results indicated that fibers improve the fracture resistance of WMA concretes. This improvement is more pronounced as the portion of opening mode (i.e., mode I) is dominant at the crack front.
- WMA concretes containing siliceous aggregates demonstrate higher fracture toughness than those containing limestone aggregates.
- Although WMA concretes are prepared at temperatures 20–55 °C lower than HMA concretes, fracture toughness of WMA is as high as that of HMA, and even some additives used in WMA mixtures provide higher fracture toughness as compared with HMA mixtures.

References

1. Hill B, Behnia B, Hakimzadeh S, Buttlar WG, Reis H (2012) Evaluation of low-temperature cracking performance of warm-mix asphalt mixtures. Transp Res Rec 2294(1):81–88
2. Frigio F, Raschia S, Steiner D, Hofko B, Canestrari F (2016) Aging effects on recycled WMA porous asphalt mixtures. Constr Build Mater 123:712–718
3. Hasan MRM, You Z, Porter D, Goh SW (2015) Laboratory moisture susceptibility evaluation of WMA under possible field conditions. Constr Build Mater 101:57–64
4. Tafti MF, Khabiri MM, Sanij HK (2016) Experimental investigation of the effect of using different aggregate types on WMA mixtures. Int J Pavement Res Technol 9(5):376–386
5. Su K, Maekawa R, Hachiya Y (2009) Laboratory evaluation of WMA mixture for use in airport pavement rehabilitation. Constr Build Mater 23(7):2709–2714
6. Fakhri M, Ghanizadeh AR, Omrani H (2013) Comparison of fatigue resistance of HMA and WMA mixtures modified by SBS. Procedia-Soc Behav Sci 104:168–177
7. Das PK, Tasdemir Y, Birgisson B (2012) Low temperature cracking performance of WMA with the use of the Superpave indirect tensile test. Constr Build Mater 30:643–649
8. Razmi A, Mirsayar M (2018) Fracture resistance of asphalt concrete modified with crumb rubber at low temperatures. Int J Pavement Res Technol 11(3):265–273
9. Mansourian A, Razmi A, Razavi M (2016) Evaluation of fracture resistance of warm mix asphalt containing jute fibers. Constr Build Mater 117:37–46
10. Pirmohammad S, Khanpour M (2020) Fracture strength of warm mix asphalt (WMA) concretes modified with crumb rubber subjected to variable temperatures. Road Mater Pavement Des (in press)
11. Hojjati Mengharpey M (2018) Investigating the effect of temperature cycling on fracture behavior of warm mix asphalt (WMA) concretes reinforced with natural fibers under mixed mode I/II loading using a new test specimen, MSc Dissertation, University of Mohaghegh Ardabili

12. Aliha M, Razmi A, Mansourian A (2017) The influence of natural and synthetic fibers on low temperature mixed mode I + II fracture behavior of warm mix asphalt (WMA) materials. Eng Fract Mech 182:322–336
13. Erdogan F, Sih G (1963) On the crack extension in plates under plane loading and transverse shear. J Basic Eng 85(4):519–525
14. Braham AF, Buttlar WG, Marasteanu MO (2007) Effect of binder type, aggregate, and mixture composition on fracture energy of hot-mix asphalt in cold climates. Transp Res Rec 2001(1):102–109
15. Lee SJ, Park J, Hong JP, Kim KW (2013) Fracture resistance of warm mix asphalt concretes at low temperatures
16. Yeon KS, Kim S, Lee HJ, Kim KW (2014) Low temperature tensile characteristics of warm mix asphalt mixtures. J Test Eval 42(4):903–911
17. Podolsky JH, Buss A, Williams RC, Cochran E (2016) Comparative performance of bio-derived/chemical additives in warm mix asphalt at low temperature. Mater Struct 49(1–2):563–575
18. Mashaan NS, Ali AH, Karim MR, Abdelaziz M (2014) A review on using crumb rubber in reinforcement of asphalt pavement. Sci World J 2014
19. Chen H, Xu Q (2010) Experimental study of fibers in stabilizing and reinforcing asphalt binder. Fuel 89(7):1616–1622
20. Roque R, Tia M, Ruth BE (1987) Asphalt rheology to define the properties of asphalt concrete mixtures and the performance of pavements. In: Asphalt rheology: relationship to mixture, ASTM International
21. Fazaeli H, Samin Y, Pirnoun A, Dabiri AS (2016) Laboratory and field evaluation of the warm fiber reinforced high performance asphalt mixtures (case study Karaj-Chaloos Road). Constr Build Mater 122:273–283
22. Kaloush KE, Biligiri KP, Zeiada WA, Rodezno MC, Reed JX (2010) Evaluation of fiber-reinforced asphalt mixtures using advanced material characterization tests. J Test Eval 38(4):400–411
23. Pirmohammad S, Hojjati Mengharpey M (2020) Influence of natural fibers on fracture strength of WMA (warm mix asphalt) concretes using a new fracture test specimen. Constr Build Mater (under review)
24. Pirmohammad S, Hojjati Mengharpey M (2018) A new mixed mode I/II fracture test specimen: numerical and experimental studies. Theor Appl Fract Mech 97:204–214
25. Hurley GC, Prowell BD (2005) Evaluation of Sasobit for use in warm mix asphalt, NCAT report 5(06)
26. Yoo MY, Jeong SH, Park JY, Kim NH, Kim KW (2011) Low-temperature fracture characteristics of selected warm-mix asphalt concretes. Transp Res Rec 2208(1):40–47
27. Singh D, Chitragar SF, Ashish PK (2017) Comparison of moisture and fracture damage resistance of hot and warm asphalt mixes containing reclaimed pavement materials. Constr Build Mater 157:1145–1153
28. Pirmohammad S, Yousefi A, Sobhi S, Vaseghi Z (2020) Fracture strength of warm mix asphalt (WMA) concretes containing reclaimed asphalt pavement (RAP). J Cent South Univ (under review)
29. Pirmohammad S (2020) Study on the effect of rejuvenation agent on the fracture behavior of warm mix asphalt (WMA) mixtures containing reclaimed asphalt pavement (RAP), University of Mohaghegh Ardabili, Research Report No. 26106

Chapter 5
Application of Nonlinear Fracture Mechanics in Asphalt Concretes

Abstract Application of nonlinear fracture mechanics for asphalt mixtures is discussed in this chapter. For this purpose, two nonlinear methods including fracture energy and J-integral are usually used to characterize the fracture resistance of asphalt mixtures. Generally, three different tests including SCB, DC(T), and SENB have been well developed in the past years to determine the fracture energy. These tests are described herein in detail. Furthermore, to calculate the critical value of J-integral, J_c, the specimens with at least two different notch lengths are required to be tested. Strain energy for each notch length is obtained by calculating the area under the load–displacement curve to the failure point. The strain energy versus the notch length is then plotted. The slope of the fitting line represents the J_c value.

5.1 Introduction

It is well known that asphalt concretes are brittle materials at low temperatures or at high loading rates while their behavior is quasi-brittle (or viscoelastic) at elevated temperatures or at low loading rates. For the cases that asphalt concretes fail in a brittle manner, linear elastic fracture mechanics (LEFM) is considered as a reasonable approach for predicting the onset of crack extension; while, in the case of viscoelastic behavior, LEFM is no longer valid due to the presence of fracture process zone (FPZ) at the tip of a crack within asphalt concrete. When the size of FPZ is much less than the dimensions of the road structure, crack growth can be described by LEFM concept; otherwise, nonlinear fracture mechanics analysis should be conducted.

One of the fracture characteristics of quasi-brittle materials is the stable crack propagation before the crack reaches its critical length, at which instable crack growth takes place. Such behavior is known as nonlinear fracture mechanism.

Fracture energy is an important parameter in nonlinear fracture mechanics. The fracture energy G_f can be defined as the amount of work to create a new surface or crack of unit length. Based on a review of nonlinear fracture mechanics, the fracture energy is one of the most appropriate parameters to describe asphalt concrete fracture. Fracture energy accounts for the large FPZ associated with quasi-brittle materials. Meanwhile, some researchers have employed another approach, namely J-integral

© Springer Nature Switzerland AG 2020

S. Pirmohammad and M. R. Ayatollahi, *Fracture Behavior of Asphalt Materials*, Structural Integrity 14, https://doi.org/10.1007/978-3-030-39974-0_5

method to measure the fracture strength of asphalt mixtures when asphalt mixtures behave in a nonlinear manner. This chapter reviews the research studies performed on the crack extension in asphalt mixture concretes exhibiting a nonlinear behavior.

5.2 Fracture Energy

Indeed, fracture energy is a measure of material resistance against thecrack propagation. Accordingly, the fracture energy G_f is defined as the amount of energy needed to propagate a crack for a unit area and is written as follows:

$$G_f = \frac{W_f}{A_{lig}}$$

(5.1)

where W_f and A_{lig} refer to the work of fracture (J) and ligament area (m^2), respectively. Thus, the fracture energy G_f is typically expressed by J/m^2. Work of fracture is the area under the load–displacement curve (see Fig. 5.1). Displacement is obtained using an extensometer measuring the load-line-displacement (LLD) or a clip gage extensometer measuring the crack mouth opening displacement ($CMOD$). During the fracture energy tests, crack propagation must be stable. Meanwhile, the ligament area is the product of ligament length and thickness of a specimen (i.e., the fracture surface). For example, for the semicircular bend (SCB) specimen, the ligament area can be obtained from the following relation:

$$A_{lig} = (r - a) \times t$$

(5.2)

where r, t, and a represent the radius of SCB, thickness of SCB, and crack length, respectively.

A great deal of efforts has been directed toward the development of fracture tests in recent years to investigate the crack initiation and propagation mechanisms.

Fig. 5.1 Load–displacement curve to calculate the work of fracture

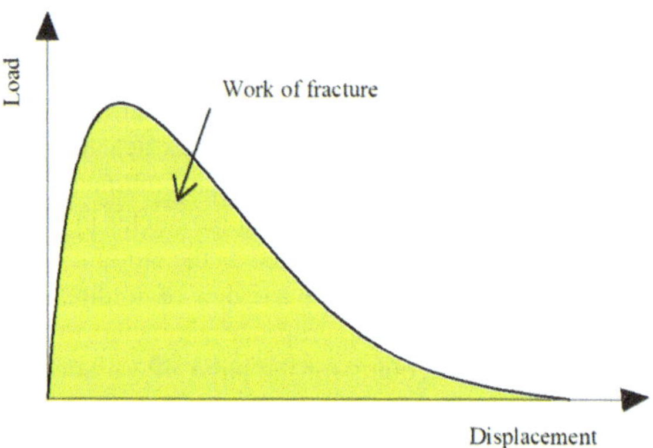

In this regard, several specimens such as SCB, DC(T), and SENB have developed to characterize the fracture energy of asphalt concretes, which are described in the subsequent sections.

5.2.1 SCB Test

The fracture test by means of the SCB is a simple and low-cost technique, which can be easily carried out using a three-point bend fixture on the semicircular samples obtained from standard cores produced by the superpave gyratory compactor (SGC) machine or taken from the field. Figure 5.2 illustrates the schematic of the SCB test. As elaborated in AASHTO TP 105-13, this test procedure needs a displacement control in two steps after inserting sitting load of $0.3 + 0.02$ kN: (i) a load of $1 + 0.1$ kN with a displacement rate of 0.06 mm/min is applied on the specimen using upper fixture. (ii) The experiment switches to *CMOD* control, and the specimen is loaded with a certain *CMOD* rate to failure [1].

Many studies can be found in the literature on characterizing fracture energy of asphalt concretes using the SCB specimen. Researchers investigated fracture energy of different asphalt mixtures at various temperatures and test conditions. Some of these researches are reviewed below.

Ozer et al. [2] conducted SCB tests under mode I loading using laboratory-prepared and compacted mixtures with varying proportions of reclaimed asphalt pavement (RAP) and recycled asphalt shingles (RAS). The tests were carried out at

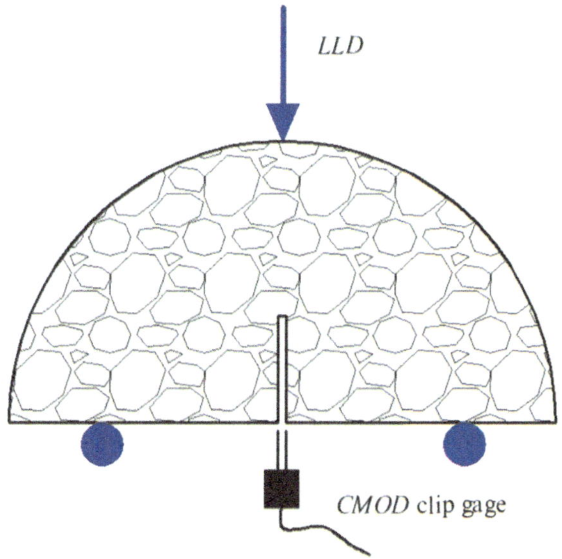

Fig. 5.2 Schematic of SCB test to obtain the fracture energy

two temperatures, 25 and -12 °C, to determine the asphalt concrete behavior at intermediate and low-temperature conditions, respectively. For the intermediate temperature, the SCB was vertically loaded by controlling the *LLD* at a rate of 50 mm/min. While, for the low temperature, the SCB test was conducted using the crack mouth opening displacement (CMOD) controlled at a rate of 1 mm/min. The fracture energy was then calculated for each mixture using the load–displacement curve recorded from the experiments. In other words, the fracture energy was obtained by finding the area under the load–displacement curve and dividing by the crack propagation area. It is pointed out that the diameter, thickness, and crack length were 150 mm, 50 mm, and 15 mm, respectively, for the SCB specimen.

Based on the results, fracture energy of the studied asphalt concretes was generally in the range of 650–750 J/m^2 for the low-temperature condition (namely -12 °C); while, it ranged from 1374 to 2307 J/m^2 for the intermediate temperature condition (namely 25 °C). Hence, the value of fracture energy decreased as the temperature reduced due to the increasing brittleness of asphalt mixtures. In addition, a consistent drop in fracture energy was observed by increasing RAP and RAS contents in the asphalt mixture. Because, brittleness of asphalt mixture increased as RAS or RAP, or both, were added to asphalt mixture.

According to Ozer et al. [3], a very tight range of fracture energy was generally seen for the low-temperature condition, even though distinctive asphalt concrete mix design characteristics (including binder type and content, RAP and RAS content and aggregate gradation and sources) were used in this study. They [3] concluded that the low-temperature fracture energy obtained from the tests was not sufficient to screen asphalt concretes with respect to their potential cracking resistance.

They also investigated the effect of loading rate by performing the tests with *LLD* rates of 6.25 mm/min and 50 mm/min at the intermediate temperature of 25 °C. Based on this study, the effect of the loading rate was more complicated. However, fracture energy values at higher loading rates produced greater fracture energy for all asphalt mixtures. It is pointed out that this result is in contrast to the results observed by other researchers, which will be discussed later.

Digital image correlation (DIC) is an optical technique to measure surface strains and displacements on a deforming specimen [4]. The DIC method was utilized in this study to assess the FPZ before and after the crack initiation to grasp the mechanisms of damage and cracking in asphalt mixtures under different temperatures and loading rates. A high-resolution charge couple device (CCD) camera system was used to collect images for the DIC optical method. Table 5.1 displays the horizontal strain field (which is the main factor for initiation and propagation of crack in the SCB tests) at peak load for the two test conditions (namely -12 °C at a *CMOD* rate of 0.7 mm/min and 25 °C at a *LLD* rate of 50 mm/min) and two asphalt concretes (namely mixtures with 0% RAS and 7% RAS). The images shown in Table 5.1 clearly display the difference in damage mechanisms at the crack front for various temperatures, loading rates, and RAS content levels. As is obvious, the damage is extended in the mastic, while the aggregates hardly exhibited any strain. This difference in strain between the binder and aggregate was also observed by Masad et al. [5] using DIC and finite element simulations and was related to the difference

Table 5.1 Horizontal strain field on the SCB (at peak load) for the two test conditions and two asphalt mixtures [3]

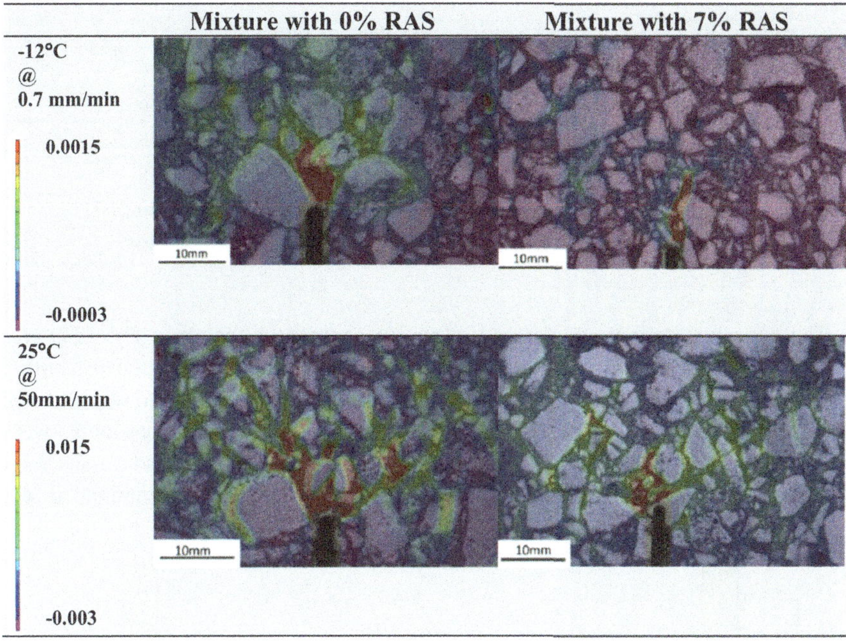

in mechanical properties of the individual elements: The aggregates were stiff while the binder was softer and more compliant; therefore, the binder strained further [6].

According to Table 5.1, by increasing the temperature, the damage was distributed over a wider area, which describes the increase in value of fracture energy for the temperature of 25 °C. While, for the case of -12 °C, the distribution of the damage was more localized, thus the failure took place more brittle. A similar observation was found for the mixtures with 0% RAS and 7% RAS, i.e., the damage localization was more pronounced for the mixture with 7% RAS tested at all conditions, which indicates a more brittle failure.

In another study, Li and Marasteanu [7] characterized fracture energy of six asphalt concretes, representing a combination of different factors like aggregate type, binder type, binder modifier, and air voids, at three low temperatures (i.e., -6, -18, and -30 °C) using the SCB specimen. They also evaluated the effect of loading rate and initial notch length on the fracture energy.

The SCB specimen of 25 mm thickness and 150 mm dimeter was used in this study. A notch with 15 mm in length and 2 mm in width was generated in the SCB specimen. In order to study the effect of the notch length on the fracture energy, the SCB specimens with two additional notch lengths of 5 and 30 mm were produced for the asphalt concrete No. 6 described in Table 5.2. Furthermore, to investigate the effect of the loading rate, they performed the tests on this mixture (i.e., AC No. 6 presented in Table 5.2) at two additional loading rates of 0.009 and 0.3 mm/min.

Table 5.2 Asphalt concrete specifications [7]

AC	Binder and modifier	Air voids		Aggregate type		Temperature (°C)
		4%	7%	Granite	Limestone	
1	PG58-28	✓		✓		−6
2	PG58-28		✓	✓		−6
3	PG64-28	✓		✓		−18
4	PG64-28	✓			✓	−18
5	PG64-28&SBS	✓		✓		−30
6	PG64-28&SBS	✓			✓	−30

The experiments were performed in an environmental chamber. The *LLD* was measured using a vertically mounted extensometer; one end was mounted on a button that was permanently fixed on a specially made frame, and the other end was attached to a metal button glued to the SCB specimen (see Fig. 5.3). The experiments were conducted at a *CMOD* rate of 0.03 mm/min, and variations of load versus *LLD* were then recorded. Before loading the specimens, they were maintained in the

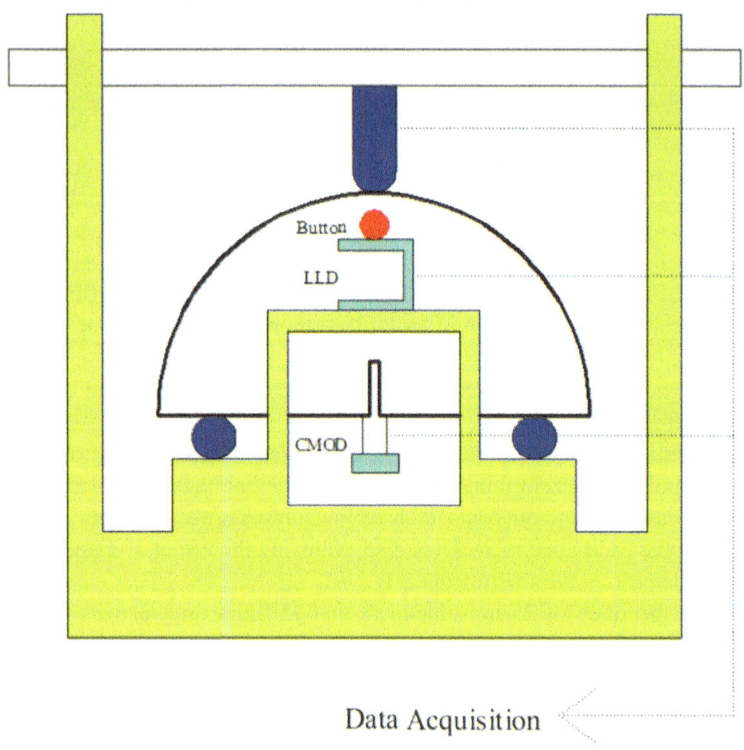

Fig. 5.3 Schematic of the SCB test setup [7]

environmental chamber at the test temperature for two hours to avoid any temperature gradient within the specimens.

According to the load–*LLD* curves obtained from the experiments, at higher temperatures (i.e., -6 °C), asphalt concretes showed more ductile behavior and have lower peak loads and larger displacements; while, they showed brittle behavior and had high peak loads and small deformation ability at the lowest temperature (i.e., -30 °C).

The results also exhibited that the asphalt mixtures containing granite aggregates had significantly higher fracture energy than those containing limestone aggregates at all the test temperatures (i.e., -6, -18, and -30 °C). In addition, the asphalt mixtures prepared with granite had higher values of peak load than those prepared with limestone at the temperatures of -30 and -18 °C; whereas, no significant difference was observed at the highest temperature (i.e., -6 °C). Hence, the aggregate type influenced the fracture resistance significantly. The same result has been found by Wu et al. [8] when studying the fracture energy of high performance concrete. Discrepancy between the results for the two aggregate types can be found via visual inspection of the fracture surfaces. According to these inspections, the crack propagated through the aggregate particles in significant portion of the fracture surface for the mixtures prepared with limestone; while, for the mixtures prepared with the granite, the crack propagated along the interface between the mastics and the aggregates. It is also found that the mixtures with 4% air void resulted in higher fracture peak load and fracture energy than those with 7% air void content, as expected. Because, more strain energy was required to break a more compacted asphalt mixture.

They also investigated the effect of the binder modifier (i.e., SBS) on the fracture energy. According to the results, the asphalt mixtures modified with SBS had obviously higher fracture energy than those with plain binder. This improvement in fracture energy was more pronounced at the two lower temperatures, such that the modified mixtures showed more than 30% increase of fracture energy compared to the unmodified mixture.

Based on the results, it was shown that the asphalt mixture containing PG58-28 binder had higher fracture energy than that containing PG64-28 binder, since the PG58-28 binder was softer than the PG64-28 binder. Hence, asphalt mixture containing PG58-28 binder was more resistant to fracture than that containing PG64-28 binder.

The results exhibited that the fracture energy reduced by increasing the loading rate at all the test temperatures. However, a significant difference among the different temperatures was observed. The fracture energy significantly changed with the loading rate at the temperature of -6 °C, while a smaller change in the fracture energy with the loading rate was found at the two lower temperatures of -30 and -18 °C.

The experiments performed on the SCB specimens with different notch lengths (5, 15, and 30 mm) showed that the fracture energy significantly differed with the notch length at -6 °C and increased as the notch length reduced. At -18 °C, the 5 mm notch length showed higher fracture energy than the 15 and 30 mm notch lengths, while the latter two exhibited no significant difference. Moreover, no significant differences in the values of fracture energy were observed at the lowest temperature of $-30\,°C$ for all the three notch lengths.

In another study, Pirmohammd et al. [9] used the SCB specimen to obtain fracture energy of asphalt mixtures. Five different asphalt mixtures were investigated in this study. Two of the mixtures contained binders with the penetration grades of 40 and 60, and the third mixture contained a binder with the penetration grade of 85 modified with 3.5% SBS (by weight of binder). The mentioned three mixtures were produced with 4% air void content. In order to investigate the effect of air void content on the fracture energy, two additional asphalt mixtures containing 7 and 10% air void were also prepared. The latter two mixtures contained the same binder with the penetration grade of 60. Meanwhile, all the mixtures had identical aggregate gradations.

The SCB specimens were manufactured by cylindrical samples of 75 mm radius and 130 mm height using superpave gyratory compactor. These samples were then cut into three disks with 32 mm in thickness using a water-cooled masonry sawing machine. The disks were then halved to produce six semicircular samples. In the next step, an initial crack with 20 mm in length was carved within each sample to produce cracked SCB specimens.

The produced cracked SCB specimens were maintained in the temperature of $-10\,°C$ for 12 h. The specimens were then loaded under pure mode I to record the load–*CMOD* curves. The value of *CMOD* was measured using a clip gage. Meanwhile, three SCB specimens were tested for each asphalt mixture, and the results were then averaged to obtain the average value of fracture energy. The results are shown in Fig. 5.4. According to this figure, the fracture energy increased as the binder penetration grade increased (or as the binder softness increased). Further improvement in fracture energy was achieved by using the SBS modified mixture. On the other hand, the fracture energy decreased as the air void content in the mixture increased. Because, air voids provide planes of weakness in the mixture, which resulted in crack initiation and propagation.

In another study, Artamendi and Khalid used the SCB test to characterize the fracture energy of different mixtures. In this study, the fracture tests were performed by

Fig. 5.4 Fracture energy of different mixtures [10]

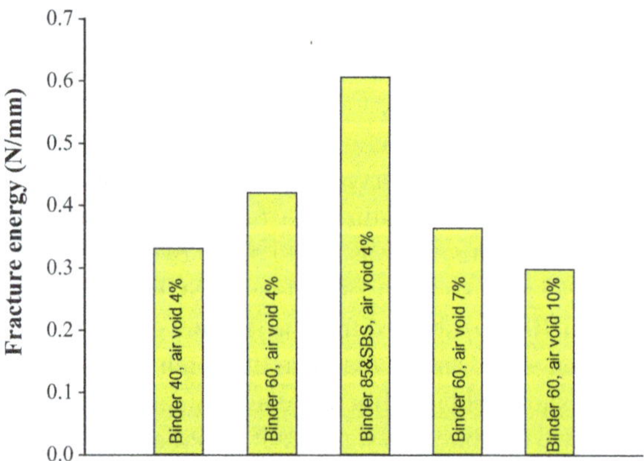

controlling the *LLD*, and finally the load–*LLD* curves recorded from the experiments were used to calculate the fracture energy. Details of this study will be discussed in Sect. 5.2.3.

5.2.2 DC(T) Test

Similar to the SCB test, the disk-shaped compact tension (DC(T)) test can be carried out by developing a crack through the specimen under displacement-controlled tensile loading. Arms are put into two openings (i.e., holes) created within the DC(T) specimen and are pulled apart with a certain rate measured by the *CMOD* (see Fig. 5.5). Unlike the SCB, loading is applied vertically, while the crack grows in the horizontal direction through the DC(T) specimen. Data is collected in the same procedure as that of the SCB test, which was described earlier.

The DC(T) test was developed by Wagoner et al. [11], and later standardized in ASTM 7313-07 "Determining fracture energy of asphalt-aggregate mixtures using the disk-shaped compacted tension geometry." According to the ASTM, DC(T) test specimens are conditioned in an environmental chamber at the test temperature for at least two hours. The DC(T) tests are recommended to be performed at 10 °C warmer than the low-temperature grade of the performance grade (PG) binder. For example, the standard test temperature of -12 °C is recommended for the PG64-22 binder [12].

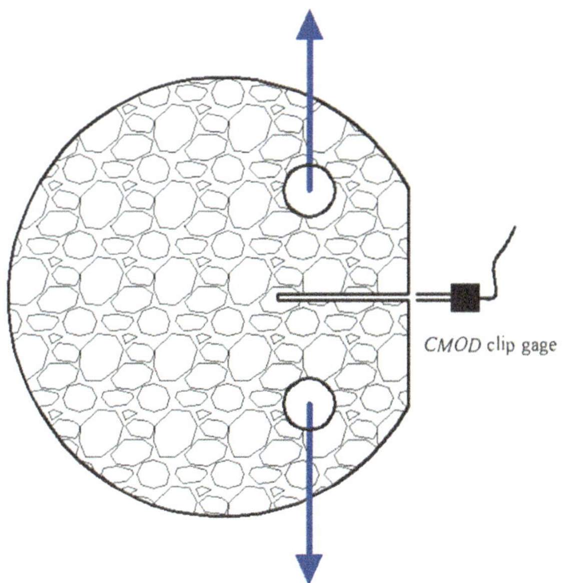

Fig. 5.5 Schematic of the DC(T) test to obtain the fracture energy

Geometry and dimensions of the DC(T) specimen recommended by Wagoner et al. [11] are shown in Fig. 5.6. They made many fracture tests on this specimen to reach the values presented in Fig. 5.6. Since the DC(T) specimen is being developed as a practical engineering fracture test, the thickness of DC(T) specimen (namely B) should reflect the thickness of asphalt concrete layer paved on the road (i.e., 25–100 mm). It is also recommended that the smallest dimension of the specimen be at least four times greater than the maximum aggregate size. For this investigation, the thickness of specimen was selected 50 mm, since the maximum aggregate size used in the mixture was 9.5 mm. Fabrication of the DC(T) specimen was nearly straightforward and can be cut from cylindrical samples to a desired thickness. The notch present in the DC(T) specimen was fabricated in two steps: (i) about half of the intended length of notch was fabricated using a water-cooled masonry saw with a notch width of 5 mm, (ii) the remained half of notch length was then created using a handsaw with a carbide-grit blade to produce a notch width of 1 mm. It was also possible to create the notch in single step using a water-cooled masonry saw with a thin blade.

An asphalt concrete mixture containing 9.5 mm maximum aggregate size and asphalt binder with PG64-22 was used for conducting fracture tests at different temperatures (i.e., -20, -10, and $0\,°C$) and loading rates (i.e., 10, 5, 1, and 0.1 mm/min) [10]. Three replicates were tested for each temperature and loading rate to calculate the average fracture energy. The fracture energy was obtained from the load–*CMOD* curve recorded from the test.

The aggregate structure influences the crack path, and thus the softening response. The crack can propagate through aggregates, between aggregate–binder interfaces, or through the binder. For each of the mentioned crack growth paths, the fracture energy would be quite different. Thus, the softening response is expected to be different because the aggregate structure for each replicate is different. According to the load–*CMOD* curves obtained from the tests, the aggregate structure did not

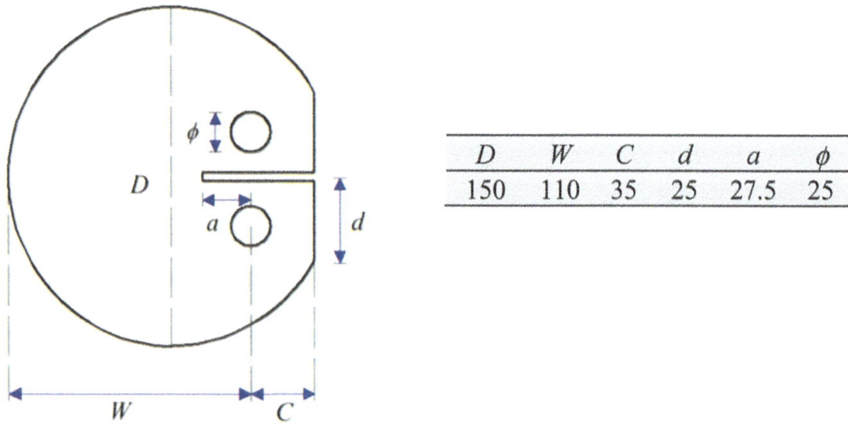

D	W	C	d	a	ϕ
150	110	35	25	27.5	25

Fig. 5.6 Geometry and dimensions of the DC(T) specimen (mm) [11]

Table 5.3 Fracture energy (J/m^2) at different temperatures and loading rates [11]

Loading rate (mm/min)	Temperature (°C)		
	−20	−10	0
10	41	276	407
5	197	318	397
1	233	328	470
0.1	245	352	848

nearly influence the load needed for crack initiation (i.e., before peak load or before softening part in the load–*CMOD* curve), while different curves were observed for the replicates after peak load (i.e., at softening part). By comparing the load–*CMOD* curves for the investigated temperatures, some significant conclusions were drawn: (i) the initial stiffness of the asphalt concrete reduced as the temperature increased, (ii) the curvature of the softening part reduced by increasing the temperature, which offered more load carrying capacity, and thus the crack propagation dissipated more energy. A similar trend was observed for the effect of loading rates. As the loading rate increased, the asphalt concrete appeared more brittle. Table 5.3 presents the average fracture energy at different temperatures and loading rates. Generally, by decreasing the temperature and increasing the loading rate, the fracture energy reduced.

In another study performed by Wagoner et al. [13], fracture energy of four different asphalt mixtures has been determined using the DC(T) specimen at three different temperatures (i.e., −20, −10, and 0 °C) and a *CMOD* rate of 1 mm/min. Furthermore, the effects of thickness and loading rate were also investigated.

Change in the thickness of the pavement layer can be reflected on the thickness of the DC(T) specimen. Hence, the effect of the DC(T) thickness on the fracture energy should be understood to compare the fracture responses between the specimens with different thicknesses. The tests were performed on the DC(T) specimens with different thicknesses ranging from 25 to 75 mm at −10 °C and loading rate of 1 mm/min. According to the results, the fracture energy increased as the specimen thickness enhanced, which agrees with the findings of Duan et al. [14]. In the rest of this study, all the tests were performed on the DC(T) specimen with a single thickness of 50 mm.

According to the results given in [13], by increasing the temperature, the fracture energy increased, which agreed with their abovementioned finding (see [11]). This result can be imputed to two reasons: (i) by increasing the temperature, the mixture became more ductile, consuming more energy for initiating and propagating a crack, (ii) the interaction between the aggregate stiffness and mastic stiffness changed by varying the temperature. At the lower temperatures, the crack tended to grow through both the aggregates and the mastic, while it propagated around the aggregates at the higher temperatures. More energy was dissipated due to aggregate bridging. The mixture containing polymer-modified binder provided a better resistance to fracture. In addition, the mixture containing softer PG58-22 binder demonstrated higher fracture energy compared to the stiffer PG64-22 binder. A similar result has been observed by Behnia et al. [15]. They obtained fracture energy of asphalt concretes containing

PG64-22 binder and PG58-28 binder both with different portions of RAP (0, 10, 20, 30, 40, and 50%) using the DC(T) specimen at $-12\,°C$ and loading rate of 1 mm/min. Based on the results, the fracture energy of the mixtures containing PG58-28 binder was greater than that of the mixture containing PG64-22 binder with different RAP contents. Furthermore, the fracture energy decreased as the portion of RAP in the mixtures containing PG58-28 binder enhanced; while, it initially increased and then decreased by increasing the portion of RAP in the mixtures containing PG64-22 binder. It is also pointed out that Li and Marasteanu [7] reported the similar trend by performing fracture tests on the SCB specimens, as mentioned earlier. According to their investigation, the asphalt mixture containing PG58-28 binder had higher fracture energy than that containing PG64-28 binder, since the PG58-28 binder was softer than the PG64-28 one.

The fracture energy calculated from the DC(T) specimen was also compared with that obtained by Wagoner et al. [16] using the SENB specimen for the same asphalt mixtures. The fracture energy calculated from these specimens was expected to be different because of differences in boundary conditions, crack front constraints, specimen size, etc. However, these specimens should rank the asphalt mixtures consistently. Generally, the value of fracture energy obtained from the DC(T) test was greater than that obtained from the SENB test.

In addition, in order to investigate the effect of loading rate on fracture response, a single asphalt mixture was tested at four different loading rate levels (10, 5, 1, and 0.1 mm/min). According to the results, the fracture energy increased as the loading rate reduced. Moreover, asphalt concrete exhibited brittle behavior without a softening part after peak load for the fastest loading rate at $-20\,°C$. The transition of the asphalt mixture from softening quasi-brittle to brittle may be attributed to the glass transition temperature. The identical phenomenon should be observed at a lower temperature and slower loading rate, if the brittle transition is attributed to the glassy transition. For this purpose, they conducted additional DC(T) tests at a temperature of $-30\,°C$ and two loading rates of 5 and 1 mm/min. Based on the results, the brittle transition of asphalt mixture depended on both the temperature and loading rate.

Buttlar et al. [12] conducted DC(T) tests to obtain fracture energy of different asphalt mixtures at the temperatures of -24, -12, and $0\,°C$. A seating load of 0.2 kN was used before starting the tests. The experiments were performed at a standard *CMOD* rate of 1 mm/min and were ended when the inserted load reduced to 0.1 kN after the peak point, as described in ASTM D7313.

The results demonstrated that the value of fracture energy increased as the test temperature increased for all the studied mixtures. In other words, the highest fracture energy was achieved at the test temperature of $0\,°C$, because the crack propagated around the aggregates, which consumed more energy compared to the low temperatures like $-24\,°C$ in which relatively straight crack growth path was observed, indicating a brittle fracture. The results also showed that fine-graded asphalt concretes were able to perform as good as or better than coarse-graded ones with similar composition at low temperatures. In addition, no differences in fracture energy were observed for the asphalt mixtures with similar binder content.

In another study, Braham et al. [17] investigated aging effects on fracture energy using the DC(T) test. Their results demonstrated that the fracture energy of asphalt mixtures increased with an increase in aging level until it reached a peak and then reduced with further aging.

5.2.3 SENB Test

Single edge notched beam (SENB) is known to be a promising test to characterize the fracture energy of asphalt materials. The size is the main factor in the selection of the SENB specimen. Furthermore, the ligament should be large enough to encompass the FPZ. For asphalt mixtures, the usual approach for selection of the specimen size is that the minimum specimen dimension should be at least three to four times larger than the maximum aggregate size to ensure that the experimental results are statistically valid [18]. Since most of the asphalt mixtures utilize a maximum aggregate size of 19 mm, the minimum ligament length of 76 mm is selected.

The SENB is also capable of performing mixed mode I/II fracture tests. The mixed mode fracture tests can be readily performed by simply offsetting the notch from the centerline of the SENB (see, e.g., [19, 20]). Table 5.4 summarizes the SENB size, test temperature, and test control used by researchers on asphalt mixtures.

The SENB test was developed by Wagoner et al. [16] to investigate fracture behavior of asphalt mixtures. They used the SENB specimen with dimensions of 375 mm in length, 100 mm in height, and 75 mm in thickness. A notch of 19 mm length was fabricated in the SENB specimen by a two-step procedure. First, half of the intended length of notch was fabricated using a water-cooled masonry saw with a notch width of 5 mm, and the remained half of notch length was then created using a handsaw with a metal cutting blade to produce a notch width of 1 mm.

Table 5.4 SENB size, test temperature, and test control used by researchers for performing fracture tests [16]

Researchers	SENB size (mm)	Test temperature (°C)	Test control
Majidzadeh et al. [21]	25 × 25 × 305 50 × 75 × 356	−5, 5, 25	*LLD* Load
Ramsamooj [22]	75 × 100 × 406	23, 9	Load
Mobasher et al. [23]	89 × 89 × 406	−1, −7	*CMOD*
Kim and El Hussein [24]	70 × 50 × 300	−5, −10, −15, −20, −25, −30	*LLD*
Bhurke et al. [25]	50.8 × 50.8 × 203.2	−10	*LLD*
Hossain et al. [26]	75 × 100 × 400	5, 25	*LLD*
Marasteanu et al. [27]	5 × 95 × 356	−18, −34	*CMOD*
Wagoner et al. [16]	100 × 75 × 375	−20, −10, 0	*CMOD*

A schematical representation of the SENB test is shown in Fig. 5.7. In the SENB test, the loading was performed by a simply supported three-point bend fixture. The fixture had a span length of 330 mm, and the rollers were 25.4 mm in diameter and were free to rotate, so they could move outward during experiment for decreasing friction. The fixture loading the centerline of SENB had a radius of 12.7 mm and can rotate in the transverse direction to apply a uniform loading across the thickness. The *CMOD* gage was attached to the SENB at the edge of the notch. Initially, the SENB test was carried out by controlling the *LLD*, but the crack growth after peak load was unstable at the lower test temperatures. The control of the test was then changed to a constant *CMOD* rate to provide a stable post-peak fracture. Meanwhile, the experiments were conducted at the temperatures of 0, −10, and −20 °C for two main reasons: to have brittle behavior and decrease viscoelastic effects, and since the low-temperature behavior of asphalt mixtures was brittle, so the control of the post-peak behavior was more difficult due to unstable crack growth. If the experiment be controllable under this condition, so it would be expected to properly conduct the experiment at higher temperatures.

The SENB specimen was kept in the cooling chamber for 3 h before performing the experiment to ensure that the temperature was uniform throughout the specimen. Then, a small preload (about 0.2 kN) was applied to the specimen to ensure that the SENB was firmly seated on the loading fixture. The experiment was then conducted using a constant *CMOD* rate of 0.7 mm/min till the load decreased to below 0.1 kN or till the specimen was completely fractured. The load–*CMOD* curve was finally recorded from the experiment.

The fracture energy was computed from Eq. (5.1) as the area under the load–*CMOD* curve normalized by the ligament and SENB thickness. The self-weight of the SENB was neglected in the fracture energy calculations at the low temperatures due to negligible change in *CMOD*. However, the self-weight may affect the test results at higher temperatures. An alternative way for decreasing the self-weight

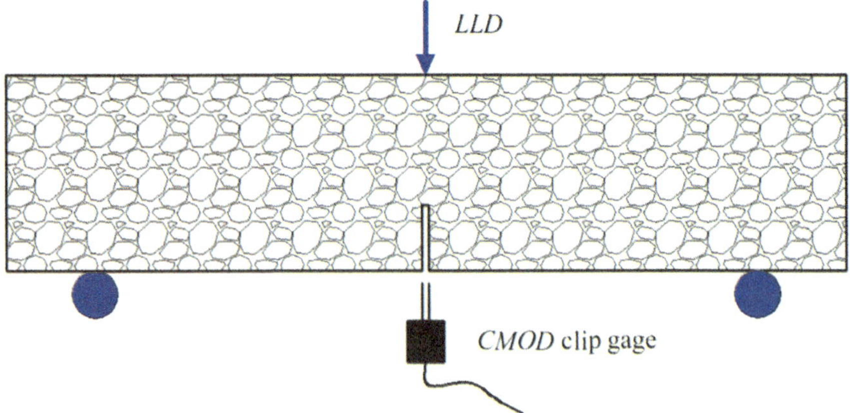

Fig. 5.7 Schematical representation of the SENB test

effect was that the specimen was inverted, such that the notch was located at the top of the SENB, and the center loading point was located at the bottom [28].

Three different asphalt mixtures were investigated in this study, and three SENB specimens were tested to assess the repeatability of this specimen. According to the load–*CMOD* curves obtained from the experiments, all the three replicates showed the same peak loads and softening curves. The overall repeatability of the SENB tests appeared to be satisfactory. Figure 5.8 shows the average values of the fracture energy obtained from the experiments for three different mixtures at the test temperatures. According to Fig. 5.8, the effect of temperature for all the mixtures was the same, i.e., the fracture energy increased as the temperature enhanced. Moreover, the mixture with 4.75 mm NMAS resulted in the highest fracture energy at all the test temperatures, while that with 19 mm NMAS produced the lowest fracture energy at all the test temperatures. This discrepancy in the fracture energy values can be attributed to two reasons: (i) the NMAS influenced the fracture energy by producing larger discontinuities in the asphalt mixture as larger aggregates were utilized, (ii) the type of binder used in the mixtures can also influence the fracture energy by having different properties (adhesion, softness, etc.). The mixture with 4.75 mm NMAS contained a polymer-modified binder, which was significantly softer than the binders used in the other mixtures.

According to Wagoner et al. [16], the DC(T) and SCB specimens, respectively, produced 35% and 50% less potential fracture area than the SENB specimen.

In another study, Artamendi and Khalid [29] calculated fracture energy of asphalt mixtures using the SENB and SCB tests. They used the SENB specimens with dimensions of 305 mm in length, 65 mm in height, and 50 mm in thickness for conducting pure mode I and mixed mode I/II experiments. A notch with 19.5 mm in length was fabricated in the SENB specimens. The notch was located in the centerline of the SENB specimens for pure mode I experiments, while for the mixed mode I/II experiments, it was created in the SENB specimen with an offset of 48.8 mm from the centerline. In addition, the span was assumed to be 244 mm in the experiments.

For the SCB tests, diameter, thickness, and crack length were taken 153 mm, 65 mm, and 23 mm, respectively. For the pure mode I, the notch was located at the

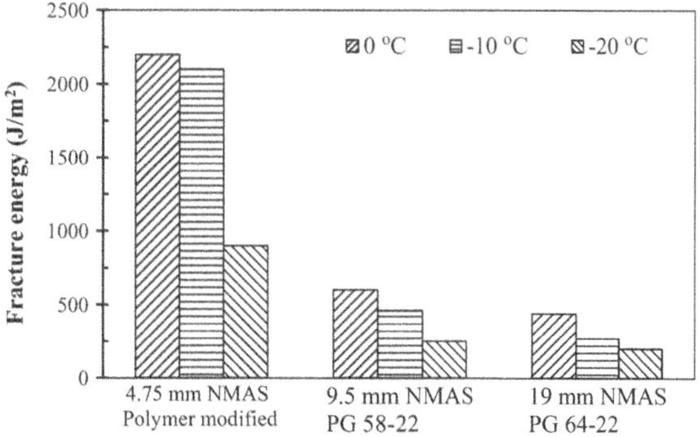

Fig. 5.8 Fracture energy obtained for the different asphalt mixtures at different temperatures [16]

centerline with an angle of 0°, while for the mixed mode I/II loading, the notch was generated at the centerline, but with an angle of 45°, with respect to the direction of the load applied to the specimen at the top of the SCB specimen.

Two types of asphalt mixtures [i.e., dense bitumen macadam (DBM) and stone mastic asphalt (SMA)] were used in this study. The DBM is a continuously graded material, where the voids produced by the interlocking aggregates are filled by smaller aggregates to provide a dense material. On the other hand, the SMA is a gap-graded aggregate structure comprising a high coarse aggregate content forming a dense aggregate skeleton, which is filled with mastic of fines, fillers, and binders. The binders with penetration grades of 50 and 100 were used in the SMA and DBM mixtures, respectively. Furthermore, binder contents were 5.5% for the SMA and 4.8% for the DBM, and the air void contents were 5.1 and 5.7% for the SMA and DBM mixtures.

A universal testing machine (UTM) with a three-point bend fixture was used to conduct the tests. The load was monotonically applied at the top of the SENB and SCB specimens by controlling the crosshead displacement (i.e., *LLD*) with a rate of 1 mm/min. The load and *LLD* were then recorded from the tests. Fracture experiments were carried out inside an environmental chamber present on the UTM. For the mode I experiments, the temperatures were selected as -10, 0, and 10 °C, while the mixed mode I/II experiments were performed only at a temperature of 0 °C. Meanwhile, three replicates were tested for each mixture, specimen and temperature condition.

By recording the load–*LLD* curves from the experiments, we can obtain the fracture energy. As shown in Fig. 5.9, fracture energies obtained from the SCB tests were more than twice those obtained from the SENB tests, while both the SCB and SENB tests were performed at the same conditions. In addition, fracture energy predominantly increased as the temperature enhanced for both specimens. The results also exhibited that the SMA mixtures had higher fracture energy than the DBM ones.

As mentioned above, the fracture energy was also obtained for the case of mixed mode I/II loading. The results are presented in Table 5.5 in which under mixed mode I/II loading, the fracture energy was higher for both the SENB and SCB specimens than under pure mode I loading. In addition, although fracture energy for the DBM mixture was less than that for the SMA mixture under mode I loading, its value for the DBM was the highest under mixed mode I/II loading.

5.3   *J*-Integral Method

The *J*-integral is a method of calculating the strain energy release rate. The *J*-integral, as a fracture mechanics parameter, was introduced by Cherepanov [30] and Rice [31], independently. In materials with significant inelastic deformation, an important part of this energy is dissipated within the plastic zone, and the rest leads to the propagation of a crack. Rice [31] showed that the *J*-integral is a path-independent line integral which can be written as follows:

Fig. 5.9 Fracture energy for the **a** SMA and **b** DBM mixtures obtained from the SENB and SCB tests at different temperatures [29]

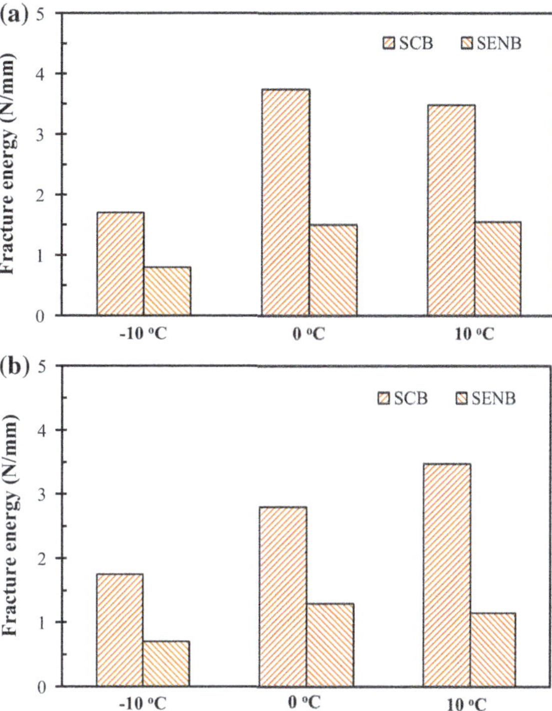

Table 5.5 Fracture energy (N/mm) of the SMA and DBM mixtures obtained from the SENB and SCB tests under pure mode I and mixed mode I/II loading at 0 °C [29]

Geometry	Material	Pure mode I	Mixed mode I/II
SCB	SMA	3.70	4.01
	DBM	2.79	5.98
SENB	SMA	1.50	2.86
	DBM	1.27	3.83

$$J = \int_{\Gamma} \left(w_s \mathrm{d}y - T_i \frac{\partial u_i}{\partial x} \mathrm{d}s \right) \tag{5.3}$$

$$w_s = \sum_{ij}^{3} \int_{0}^{\varepsilon_{ij}} \sigma_{ij} \mathrm{d}\varepsilon_{ij} \tag{5.4}$$

$$T_i = \sum_{ij}^{3} \sigma_{ij} n_{\Gamma j} \tag{5.5}$$

where w_s and T_i are the strain energy density and the traction vector, respectively. In addition, Γ is an arbitrary contour around the crack tip, n_Γ is the unit vector normal

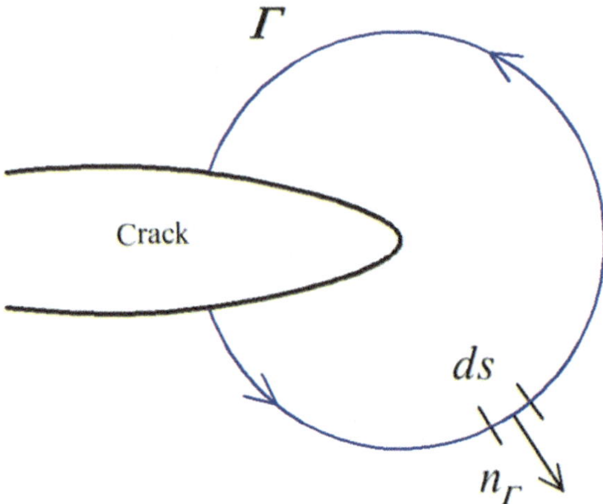

Fig. 5.10 Definition of a contour around the crack tip

to Γ (see Fig. 5.10), and σ, ε, and u are the stress, strain, and displacement fields, respectively [32].

Later, experimental methods were developed to measure the critical fracture properties using laboratory specimens. The J_{Ic} is defined as the point at which large-scale plastic yielding during propagation occurs under mode I loading [33].

On the asphalt mixtures, many researches can be found in the literature to evaluate their fracture property using the J-integral method (see, e.g., [34–38]). Some of these investigations are discussed herein.

Fracture resistance of asphalt mixtures can be also characterized by the critical strain energy release rate (i.e., the critical value of J-integral, or J_c), which is written as follows:

$$J_c = \left(\frac{U_1}{t_1} - \frac{U_2}{t_2} \right) \frac{1}{a_2 - a_1} \tag{5.6}$$

where U, a, and t are the strain energy to failure, notch length, and specimen thickness, respectively. It is worth noting that asphalt mixtures with greater J_c values represent better fracture resistance. In order to calculate the critical value of J-integral J_c, the specimens with at least two notch lengths are required to be tested. Strain energy (i.e., U) for each notch length is obtained by calculating the area under the load–displacement curve to the failure point.

Cooper et al. [39] obtained the fracture resistance of asphalt concretes using the SCB specimen. In this study, three notch lengths of 25.4, 31.8, and 38 mm were fabricated in the SCB specimens at the centerline. The experiments were performed at the test temperature of 25 °C. The SCB specimens were monotonically loaded until failure with a constant crosshead displacement rate of 0.5 mm/min using a three-point bend fixture. The load–LLD curves were then recorded from the experiments to calculate the critical value of J-integral, J_c from the Eq. (5.6). Their results showed that the asphalt mixtures containing an elastomeric type of polymer-modified binder had the highest value of J_c.

They [40] also evaluated fracture resistance of asphalt mixtures containing bio-binder (i.e., green asphalt mixtures). The bio-binder employed in their study was a tall oil derived from pine trees. Based on the results given in [40], the warm mix asphalt (WMA) concretes modified with bio-binder possessed stiffer properties than the conventional WMA concretes. This was because the use of bio-binder enhanced the brittleness of the binder at the intermediate temperature. Furthermore, the magnitude of J_c increased with an increase in the performance grade.

Mahmoud et al. [41] used the SCB test and J_c concept to evaluate the fracture resistance of asphalt mixtures. The SCB specimens with three notch lengths of 25.4, 31.8, and 38 mm were used to conduct the tests at 20 °C. The diameter, thickness, and span of the SCB specimens were 152 mm, 57 mm, and 127 mm, respectively. The experiments were performed with a constant crosshead deformation rate of 0.508 mm/min using a three-point bend fixture. As mentioned earlier, the strain energy U for each notch length was obtained by computing the area under the load–displacement curve to the failure point. Figure 5.11 displays the strain energy at different notch lengths. The strain energy reduced as the notch length increased. The slope of the fitting line represents the critical *J*-integral (i.e., J_c) value.

Wang et al. [42] conducted the SCB tests to find the value of J_c. The SCB specimens with three different notch lengths of 15, 20, and 25 mm were selected and were loaded with a constant crosshead deformation rate of 50 mm/min using a three-point bend fixture at −15 °C. The diameter d and thickness t of the SCB specimen were 150 and 25 mm, and the span value during the tests was assumed to be $0.8d$. Figure 5.12 shows the load–*LLD* curves obtained from the tests. The strain energies were calculated as 4055.799, 3105.264, and 1571.303 N mm for the SCB specimens with notch lengths of 15, 20, and 25 mm, respectively. The critical *J*-integral (i.e., J_c value) was then calculated from the relative curve between the strain energy per unit thickness of the corresponding notch length, and the least square method to fit the curve, as displayed in Fig. 5.13. As mentioned above, the slope of the linear fitting

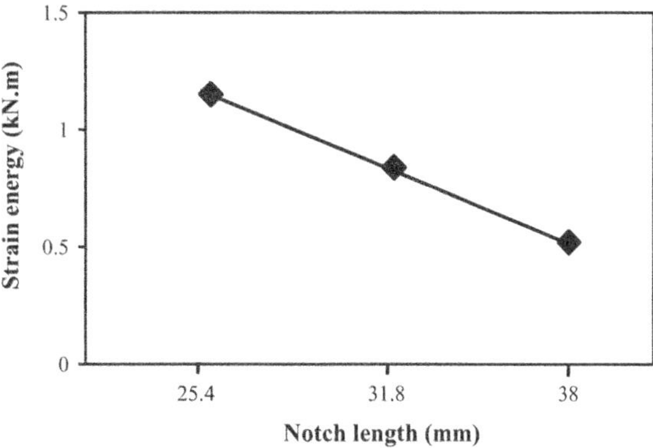

Fig. 5.11 Strain energies U calculated from the load–*LLD* curves at different notch lengths [41]

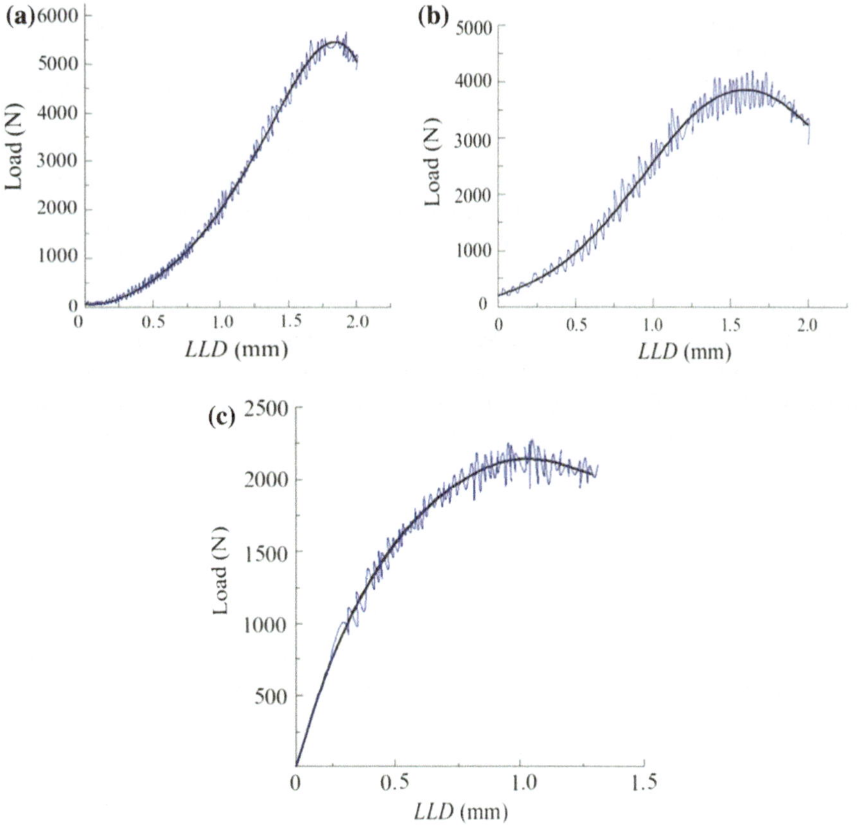

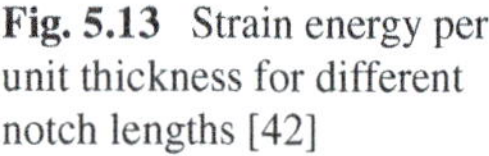

Fig. 5.12 Load–*LLD* curves obtained from the SCB tests with different notch lengths of **a** 15 mm, **b** 20 mm, **c** 25 mm [42]

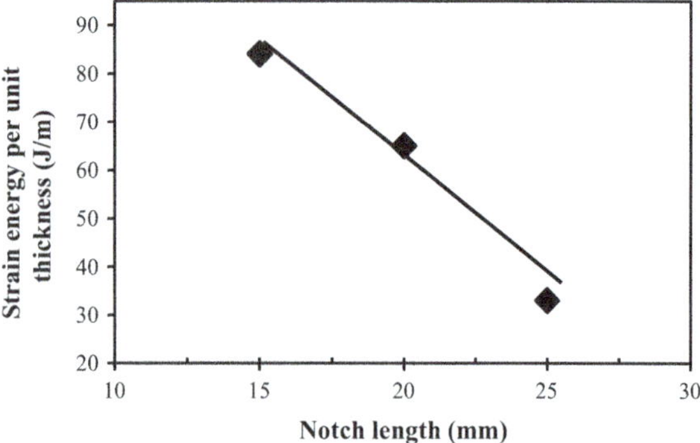

Fig. 5.13 Strain energy per unit thickness for different notch lengths [42]

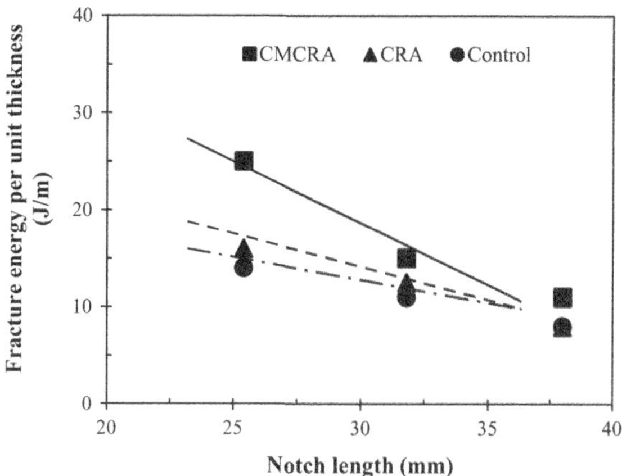

Fig. 5.14 Strain energy per unit thickness for different notch lengths and different mixtures [43]

line represented the value of J_c for the rubber-modified asphalt mixture, which was obtained as 4.97 kJ/m^2.

In another study, Mull et al. [43] used the SCB specimen with the same dimensions and loading conditions employed by Mahmoud et al. [41], as mentioned above. In this study, asphalt mixtures with three different binders were investigated: (i) control binder (with PG70-28) produced from air-blown asphalt with no catalyst, called control, (ii) binder (with PG70-22) modified by plain crumb rubber, called CRA, and (iii) chemically modified crumb rubber binder (with PG76-28), called CMCRA. The CRA and CMCRA contained the same rubber content. The SCB tests including three different notch lengths of 25.4, 31.8, and 38 mm were performed with a crosshead speed of 0.5 mm/min at 24 °C. Figure 5.14 shows the results of strain energy per thickness for different notch lengths and different mixtures. According to this figure, the mixture containing CMCRA exhibited the highest fracture resistance [i.e., the highest slope of the fitting line (or J_c)], while the mixture containing the control binder showed the lowest J_c. The higher fracture resistance of the mixture containing CMCRA was attributed to the chemical modification, which caused the mixture to become more resistant to fracture.

Bhurke et al. [25] used the SENB test to characterize fracture resistance of different asphalt mixtures. A minimum of four different notch lengths ranging from 40 to 60% of the SENB height (i.e., 20.32, 22.86, 25.40, 27.94, and 30.48 mm) was fabricated in the SENB specimens. The SENB specimens with dimensions of 203.2 mm × 50.8 mm × 50.8 mm were loaded with a constant crosshead displacement rate of 1.27 mm/min on a 177.8 mm span using a three-point bend fixture at the temperature of −10 °C. Load–*LLD* curves were recorded from the experiments. The strain energy per unit thickness (i.e., U/t) was then plotted versus the notch length, and a linear fit to these points was obtained through regression. The slope of this line gave the value of J_c. According to the results, the styrene ethylene butylene styrene (SEBS) modified asphalt mixture exhibited better fracture resistance than the unmodified one. The highest improvement was achieved by using 3% SEBS (among the 0, 3 and 5% SEBS). However, further increase in the use of SEBS worsened the fracture resistance. Likewise, the Elvaloy (epoxy-terminated reacting polyolefin) modified asphalt mixture did not exhibit any improvement in the fracture resistance. They used three different dosages of 0, 1, and 2% ELVALOY in this investigation.

In another study, Dongre et al. [44] used the SENB test for obtaining fracture resistance of different asphalt mixtures at different test temperatures of -20.6, -12.2, -3.9, and 4.4 °C. All of the asphalt mixtures exhibited higher values of critical J-integral as the temperature increased. At the higher temperatures, the softer asphalts generally demonstrated greater J_c values.

5.3.1 Summary

In this chapter, two different methods including fracture energy and J-integral were discussed to characterize the fracture resistance of asphalt mixtures.

Generally, three different tests including SCB, DC(T), and SENB have been well developed in the past years to obtain the fracture energy. Researchers investigated fracture resistance of various asphalt mixtures at different environmental and loading conditions. The important results are summarized herein, as follows:

- Fracture energy increases as the temperature increases.
- Stiffer aggregates used in asphalt mixtures show higher value of fracture energy. For example, asphalt mixtures containing granite aggregates have significantly higher fracture energy than those containing limestone aggregates.
- Fracture energy increases with a decrease in the air void content.
- Modified asphalt mixtures exhibit higher value of fracture energy compared to plain asphalt mixtures. This improvement in fracture energy is more pronounced at lower temperatures.
- Fracture energy increases as softer binder is used in asphalt mixture.
- Fracture energy decreases with increase in the loading rate.
- Fracture energy increases as the specimen thickness increases.
- Fracture energy decreases as larger aggregates are used in the mixture.
- Different specimens show different values of fracture energy.
- Fracture energy obtained from the mixed mod I/II test is higher than that obtained from the pure mode I test.

Critical J-integral was another fracture resistance indicator which was widely used by researchers to evaluate fracture behavior of asphalt mixtures. In order to calculate the critical value of J-integral J_c, the specimens with at least two notch lengths were required to be tested. Strain energy for each notch length was obtained by calculating the area under the load–displacement curve to the failure point. The strain energy versus the notch length was then plotted. The slope of the fitting line represented the critical J-integral J_c value, indicating that the strain energy reduced as the notch length increased.

References

1. El Khatib AK (2016) Development of a performance-based test method for quantification of cracking potential in asphalt pavement materials. Thesis for M.Sc. degree, University of Illinois at Urbana-Champaign
2. Ozer H, Al-Qadi IL, Singhvi P, Khan T, Rivera-Perez J, El-Khatib A (2016) Fracture characterization of asphalt mixtures with high recycled content using Illinois semicircular bending test method and flexibility index. Transp Res Rec J Transp Res Board 2575:130-137
3. Ozer H, Al-Qadi IL, Lambros J, El-Khatib A, Singhvi P, Doll B (2016) Development of the fracture-based flexibility index for asphalt concrete cracking potential using modified semicircle bending test parameters. Constr Build Mater 115:390–401
4. Sutton MA, Orteu JJ, Schreier H (2009) Image correlation for shape, motion and deformation measurements: basic concepts, theory and applications. Springer Science & Business Media
5. Masad E, Somadevan N, Bahia HU, Kose S (2001) Modeling and experimental measurements of strain distribution in asphalt mixes. J Transp Eng 127(6):477–485
6. Al-Qadi IL, Ozer H, Lambros J, El Khatib A, Singhvi P, Khan T, Rivera-Perez J, Doll B (2015) Testing protocols to ensure performance of high asphalt binder replacement mixes using RAP and RAS. Illinois Center for Transportation/Illinois Department of Transportation
7. Li XJ, Marasteanu M (2010) Using semi circular bending test to evaluate low temperature fracture resistance for asphalt concrete. Exp Mech 50(7):867–876
8. Wu K-R, Chen B, Yao W, Zhang D (2001) Effect of coarse aggregate type on mechanical properties of high-performance concrete. Cem Concr Res 31(10):1421–1425
9. Pirmohammad S, Khoramishad H, Ayatollahi M (2015) Effects of asphalt concrete characteristics on cohesive zone model parameters of hot mix asphalt mixtures. Can J Civ Eng 43(3):226–232
10. Pirmohammad S (2013) Study of mixed mode I/II fracture behavior of asphaltic materials at low temperatures. Ph.D. dissertation, Mechanical engineering, Iran University of Science and Technology
11. Wagoner M, Buttlar WG, Paulino G (2005) Disk-shaped compact tension test for asphalt concrete fracture. Exp Mech 45(3):270–277
12. Buttlar W, Na Chiangmai C, Al-Qadi IL, Murphy TR, Pine WJ (2015) Designing, producing, and constructing fine-graded hot mix asphalt on Illinois Roadways. Illinois Center for Transportation/Illinois Department of Transportation
13. Wagoner M, Buttlar W, Paulino G, Blankenship P (1929) Investigation of the fracture resistance of hot-mix asphalt concrete using a disk-shaped compact tension test. Transp Res Rec J Transp Res Board 2005:183–192
14. Duan K, Hu X-Z, Wittmann FH (2003) Thickness effect on fracture energy of cementitious materials. Cem Concr Res 33(4):499–507
15. Behnia B, Dave E, Ahmed S, Buttlar W, Reis H (2011) Effects of recycled asphalt pavement amounts on low-temperature cracking performance of asphalt mixtures using acoustic emissions. Transp Res Rec J Transp Res Board 2208:64–71
16. Wagoner MP, Buttlar WG, Paulino GH (2005) Development of a single-edge notched beam test for asphalt concrete mixtures. J Test Eval 33(6):452–460
17. Braham AF, Buttlar WG, Clyne TR, Marasteanu MO, Turos MI (2009) The effect of long-term laboratory aging on asphalt concrete fracture energy. Asphalt Paving Technology 2009, AAPT
18. Romero P, Masad E (2001) Relationship between the representative volume element and mechanical properties of asphalt concrete. J Mater Civ Eng 13(1):77–84
19. Song SH, Paulino GH, Buttlar WG (2006) Simulation of crack propagation in asphalt concrete using an intrinsic cohesive zone model. J Eng Mech 132(11):1215–1223
20. Song SH, Paulino GH, Buttlar WG (2006) A bilinear cohesive zone model tailored for fracture of asphalt concrete considering viscoelastic bulk material. Eng Fract Mech 73(18):2829–2848
21. Majidzadeh K, Kauffmann E, Ramsamooj D (1971) Application of fracture mechanics in the analysis of pavement fatigue. Association of Asphalt Paving Technologists Proc

22. Ramsamooj D (1991) Prediction of fatigue life of asphalt concrete beams from fracture tests. J Test Eval 19(3):231–239
23. Mobasher B, Mamlouk MS, Lin HM (1997) Evaluation of crack propagation properties of asphalt mixtures. J Transp Eng 123(5):405–413
24. Kim KW, El Hussein M (1997) Variation of fracture toughness of asphalt concrete under low temperatures. Constr Build Mater 11(7–8):403–411
25. Bhurke A, Shin E, Drzal L (1997) Fracture morphology and fracture toughness measurement of polymer-modified asphalt concrete. Transp Res Rec J Transp Res Board 1590:23–33
26. Hossain M, Swartz S, Hoque E (1999) Fracture and tensile characteristics of asphalt-rubber concrete. J Mater Civ Eng 11(4):287–294
27. Marasteanu MO, Dai S, Labuz JF, Li X (2002) Determining the low-temperature fracture toughness of asphalt mixtures. Transp Res Rec 1789(1):191–199
28. Carpinteri A, Shah S (2014) Fracture mechanics test methods for concrete. CRC Press
29. Artamendi I, Khalid HA (2006) A comparison between beam and semi-circular bending fracture tests for asphalt. Road Mater Pavement Des 7(sup1):163–180
30. Cherepanov G (1967) Crack propagation in a continuum. Prikl Mat Mekh 31(3):476–488
31. Rice JR (1968) A path independent integral and the approximate analysis of strain concentration by notches and cracks. J Appl Mech 35(2):379–386
32. Llopart Prieto L (2007) Modelling and analysis of crack turning on aeronautical structures
33. Chawla KK, Meyers M (1999) Mechanical behavior of materials. Prentice Hall Upper Saddle River
34. Shu X, Huang B, Vukosavljevic D (2010) Evaluation of cracking resistance of recycled asphalt mixture using semi-circular bending test. Paving materials and pavement analysis, pp 58–65
35. Zhao S, Huang B, Shu X, Ye P (2014) Laboratory investigation of biochar-modified asphalt mixture. Transp Res Rec J Transp Res Board 2445:56–63
36. Liu JH (2011) Low temperature cracking evaluation of asphalt rubber mixtures using semi-circular bending test. Adv Mater Res Trans Tech Publ, pp 4201–4206
37. Cooper SB III, Mohammad LN, Elseifi MA (2011) Laboratory performance characteristics of sulfur-modified warm-mix asphalt. J Mater Civ Eng 23(9):1338–1345
38. Mohammad LN, Cooper SB Jr, Elseifi MA (2011) Characterization of HMA mixtures containing high reclaimed asphalt pavement content with crumb rubber additives. J Mater Civ Eng 23(11):1560–1568
39. Cooper SB III, Mohammad LN, Kabir S, King W Jr (2014) Balanced Asphalt mixture design through specification modification: Louisiana's experience. Transp Res Rec 2447(1):92–100
40. Mohammad L, Elseifi M, Cooper S, Challa H, Naidoo P (2013) Laboratory evaluation of asphalt mixtures that contain biobinder technologies. Transp Res Rec J Transp Res Board 2371:58–65
41. Mahmoud E, Saadeh S, Hakimelahi H, Harvey J (2014) Extended finite-element modelling of asphalt mixtures fracture properties using the semi-circular bending test. Road Mater Pavement Des 15(1):153–166
42. Wang H, Zhang C, Yang L, You Z (2013) Study on the rubber-modified asphalt mixtures' cracking propagation using the extended finite element method. Constr Build Mater 47:223–230
43. Mull M, Stuart K, Yehia A (2002) Fracture resistance characterization of chemically modified crumb rubber asphalt pavement. J Mater Sci 37(3):557–566
44. Dongre R, Sharma M, Anderson D (1989) Development of fracture criterion for asphalt mixes at low temperatures. Transp Res Rec 1228:94–105

Chapter 6
Summary

Abstract This chapter presents a summary of the chapters discussed in this book. Particularly, the results of research studies dealing with the effects of different parameters such as mode of loading, temperature, additives, and fiber on the fracture behavior of asphalt mixtures are presented.

6.1 Introduction

In the Chap. 1 of this book, an introduction to asphalt concretes was given. For example, ingredients of asphalt concretes including aggregates, binder, and modifiers were briefly described. Then, different modes of deterioration in asphalt concretes including cracking, surface deformation, disintegration, and surface defects were introduced. Furthermore, all types of cracks appearing in asphalt pavements such as fatigue (or Alligator) cracking, longitudinal cracking, transverse cracking, block cracking, slippage cracking, reflective cracking, and edge cracking were introduced.

Chapter 2 dealt with the computational modeling of asphalt concretes using different methodologies such as discrete element method (DEM), finite element method (FEM), and extended finite element method (XFEM). In the DEM, particles are bonded together at contact points and are separated by external forces. Two homogenous and heterogeneous discrete element models have been generally developed to model crack growth behavior of asphalt concretes. In the homogenous models, behavior of material is considered cohesive at the crack trajectory whilst elastic behavior is assumed at other regions. In the heterogeneous discrete element models, an image processing technique is employed to capture the microstructure of materials. Digitalized specimen image with different aggregate size is obtained by scanning laboratory asphalt specimens. Structure of asphalt concrete mixtures is made up of three different phases including aggregates, mastic (binder + fine sands), and interface between aggregates and mastic. Material properties (i.e., elastic modulus, tensile strength, and fracture energy) of these phases are required for numerical simulations, and must be determined separately. Aggregates are assumed to behave in linear elastic without softening. The mastic and interface behave in nonlinear elastic with cohesive softening and adhesive softening, respectively. Tensile strength of asphalt concrete is

© Springer Nature Switzerland AG 2020

S. Pirmohammad and M. R. Ayatollahi, *Fracture Behavior of Asphalt Materials*, Structural Integrity 14, https://doi.org/10.1007/978-3-030-39974-0_6

characterized from indirect tension test (IDT), and fracture energy is obtained from one of the three DC(T), SENB, and SCB tests.

In the subsequent section, the effects of various parameters including vehicle wheel position, horizontal load, elasticity, and thickness of the road layers on the crack tip parameters (i.e., stress intensity factors, K_I, K_{II}, and T-stress) were investigated using two-dimensional (2D) FEM. For this purpose, a four-layer road structure including asphalt concrete, base, sub-base, and sub-grade layers was modeled in ABAQUS. Meanwhile, a top-down crack was regarded in the middle of the asphalt concrete layer. The results exhibited that the regions surrounding a top-down crack may be classified into three zones:

(i)   First zone, which corresponds to regions farther from crack, and only pure mode I deformation is observed at crack tip.
(ii) Second zone, which corresponds to regions very close to crack, and only pure mode II deformation is observed at crack tip.
(iii) Third zone, which corresponds to regions between the abovementioned zones, and mixed I/II deformation mode takes place at crack tip.

According to the results, the horizontal load induced from vehicle acceleration or braking enhanced the crack parameters (i.e., K_I, K_{II}, and T-stress) significantly. Thus, a cracked asphalt pavement is exposed to a critical loading condition where vehicles brake or accelerate (e.g., at the intersections). The following results were also obtained from the finite element analyses of the abovementioned cracked asphalt pavement:

- Crack tip parameters increase as elasticity modulus of asphalt concrete enhances. Moreover, further reduction in K_I is achieved by increasing the elasticity modulus of beneath layers. The change in the elasticity modulus of sub-grade layer does not influence K_{II} and T-stress. Hence, in order to avoid crack propagation at intersections, bus stations, etc., a material with low elasticity modulus should be performed at the top layer of road structure (e.g., modified asphalt concrete mixtures), and a material with high elasticity modulus at beneath layers including base, sub-base, and sub-grade.
- Crack tip parameters decline as thickness of road layers increases. Generally, K_I and K_{II} are more influenced by asphalt concrete thickness than thickness of base and sub-base layers.

Three-dimensional FEM can result in more accurate and reliable results as compared with two-dimensional FEM because 2D finite element modeling of road structure does not consider significant aspects of crack growth such as mode III deformations. Three-dimensional finite element modeling of a cracked road structure exhibited that mode I deformation is a major threat for the crack propagation and thereby deterioration of road structures. Other than the opening mode (i.e., mode I), the in-plane and out-of-plane shear deformation modes (i.e., mode II and mode III) were also observed at the crack front. As vehicle wheels were located far from crack plane, the shear deformations were negligible compared to the total deformations available at the crack front. The shear deformations became dominant relative to opening deformation mode by approaching vehicle wheels to crack plane, and as vehicle wheels were located very close to the crack plane, the total deformation was nearly related to the shear modes.

Extended finite element method (XFEM) has been proved to be a very efficient computational method to characterize the discontinuous mechanical problems such as crack extensions. In the XFEM, the numerical model is divided into two regions. In the first region, the classical finite element meshes are generated for the un-cracked part of geometry; while, in the second region, the meshes defined in the cracked part of geometry are enriched by appropriate functions. Thus, the XFEM incorporates enrichment functions to solve fracture problems.

In the simulation of crack propagation using cohesive zone model (CZM), four material properties including fracture energy, tensile strength (for the cohesive elements), Young's modulus, and Poisson's ratio (for the bulk material) together with the load–CMOD curve should be experimentally determined. These parameters are defined in the numerical simulations as input data, and the load–CMOD curve is obtained. Finally, the fracture energy and tensile strength incorporated into the finite element simulation are calibrated by fitting the numerically and experimentally obtained load–CMOD curves to find the cohesive parameters including the cohesive energy and the cohesive strength.

Chapter 3 discussed on the fracture behavior of HMA concretes. Different test specimens including the SENB, DC(T), SCB, etc., have been employed by researchers to measure the fracture toughness of HMA concretes. However, the SCB specimen is mostly employed for this purpose. The procedure of performing fracture tests using the mentioned specimens was also described in this chapter.

The effects of different parameters including aggregate type, aggregate gradation, air void content, binder, temperature, nanomaterials, fibers, and additives on the fracture toughness of HMA mixtures were discussed by reviewing investigations performed on these issues. The following trends for the effects of the mentioned parameters on fracture toughness have been often reported by researchers.

- Granite aggregates perform better than limestone ones at low temperatures, whereas this trend is reversed at high temperatures. Based on the results, the granite aggregates are not fractured at either low or high temperatures, while the limestone aggregates are only fractured at low temperatures, whereas at high temperatures, they remain intact.
- The use of coarser aggregates in the preparation of HMA mixtures provides higher fracture resistance. As the size of aggregates in the mixture reduces, the risk of initiating micro-cracks inside the wider space of filler and binder part of asphalt mixture increases, making the mixture more vulnerable to failure.
- Fracture toughness reduces as air void content of HMA mixture increases. More compacted mixture has higher density, and a dense mixture requires more load and strain energy for crack extension. In addition, air voids in HMA mixture are sources of stress concentration; hence, they can accelerate the crack extension at low temperatures.

- At low and mid temperatures, the addition of binder does not affect fracture energy due to creating a brittle matrix at low temperatures, whereas at high temperatures, the addition of 0.5% binder increases the fracture energy significantly because the binder can dissipate tremendous amounts of energy. Hence, more binder content does not necessarily increase the fracture energy at low temperatures but does increase the fracture energy at high temperatures.
- Binder with higher penetration grade leads to increase in the fracture toughness of HMA concrete. By increasing the penetration grade of a binder, the softness of HMA mixture increases. This increase in the softness is responsible for improving the fracture toughness at low temperatures.
- As temperature goes down, the fracture toughness of HMA concretes increases due to the increase of binder stiffness, and further decrease in temperature leads to reduction in the fracture toughness due to the differential thermal contraction (DTC) damage. In addition, temperature variations result in reduction of fracture toughness. The DTC damage can be explained such that as temperature decreases, both aggregate and binder contract but the binder surrounding the aggregate particle contracts more than the aggregate because of the significant difference in the coefficients of thermal expansion of aggregate and binder. Hence, the tensile stresses developed within the binder lead to formation of hairline (micro)-cracks which results in weakening the asphalt concrete.
- Fracture resistance of HMA mixtures initially decreases and then increases as the portion of mode II relative to mode I enhances at the crack front; as a result, HMA mixtures have their minimum fracture resistance as they are subjected to mixed mode I/II loading.
- The fracture surfaces of SCB specimens subjected to different modes of loading exhibit that for pure mode I loading, crack initiates along the initial crack line, and develops straightly toward the top fixture whilst for the mixed mode I/II and pure mode II loadings, crack kinks from the initial crack line, and propagates along a curvilinear path. This manner of the crack growth path is attributed to the maximum tensile stress around the crack front, which is not anymore along the initial crack when the SCB specimen is loaded under mixed mode I/II or pure mode II. Meanwhile, as the proportion of shear mode (i.e., mode II) at the crack front of the SCB specimen increases, the initiation angle of fracture becomes greater.
- Different types of nanomaterials such as Nano Fe_2O_3, CNTs (carbon nanotubes), Nano Al_2O_3, Nano silica, and nanoclays have been added to HMA mixtures to investigate their effect on the fracture toughness of HMA concretes under different modes of loading. The results showed that nanomaterials can significantly improve the fracture toughness of HMA mixtures.
- Likewise, different types of fibers such as kenaf, goat wool, basalt, and carbon have been incorporated into HMA mixtures to study their effect on the fracture toughness of HMA concretes under different modes of loading. The results demonstrated that fibers can significantly increase the fracture resistance of HMA mixtures. Because, fibers provide additional tensile strength, and, therefore, improve the fracture strength of asphalt mixtures.

- Additives have also positive effect on the fracture resistance of asphalt mixtures. For example, SBS improves the fracture resistance of HMA mixture at any temperature and loading conditions. Interestingly, the effect of SBS is more pronounced as HMA mixture is exposed to lower temperatures. Indeed, at low temperatures, the SBS imparts increased softness to the brittle binder, and also provides improved adhesion between aggregates and binder. The results also revealed that crumb rubber can improve the fracture toughness due to imparting ductility to the base binder and improving the adhesion between binder and aggregates.

Fracture behavior of HMA concretes subjected to mixed mode I/III and pure mode III loadings was finally evaluated in Chap. 3. The results revealed that the presence of mode III can aggravate the fracture resistance of HMA mixtures significantly.

Chapter 4 discussed on the fracture behavior of WMA concretes. The effects of different parameters including mode of loading, temperature, crumb rubber, fiber, and aggregate type on fracture resistance of WMA mixtures were reviewed. The following trends for the effects of the mentioned parameters on fracture toughness have been often observed by researchers:

- Fracture resistance of WMA concretes decreases as the amount of shear load at crack front increases, and minimum resistance of WMA concretes to fracture is achieved under a mixed mode I/II loading not under a pure mode (i.e., I or II).
- As temperature drops, fracture resistance of WMA concretes increases due to the increase of binder stiffness, and further decrease in temperature results in reduction of fracture toughness due to the DTC damage. Furthermore, temperature variations lead to reduction in fracture resistance of WMA concrete.
- The use of crumb rubber has a positive effect on the fracture toughness of WMA concrete.
- Different types of fibers such as jute, FORTA, kenaf, and basalt were incorporated to WMA mixture, and the results indicated that fibers improve the fracture resistance of WMA concretes. This improvement is more pronounced as the portion of opening mode (i.e., mode I) is dominant at the crack front.
- WMA concretes containing siliceous aggregates demonstrate higher fracture toughness than those containing limestone aggregates.
- Although WMA concretes are prepared at temperatures 20–55 °C lower than HMA concretes, but fracture toughness of WMA is as high as that of HMA, and even some additives used in WMA mixtures provide higher fracture toughness as compared with HMA mixtures.

Chapter 5 discussed on the application of nonlinear fracture mechanics for asphalt mixtures. For this purpose, two nonlinear methods including fracture energy and J-integral are usually used to characterize the fracture resistance of asphalt mixtures. Generally, three different tests including SCB, DC(T), and SENB have been well developed in the past years to determine the fracture energy. These tests have been described in Chap. 5 with more details. Researchers investigated the fracture resistance of various asphalt mixtures at different environmental and loading conditions. The following results have been reported by researchers:

- Fracture energy increases as temperature increases.
- Stiffer aggregates used in asphalt mixtures show higher value of fracture energy. For example, asphalt mixtures containing granite aggregates have significantly higher fracture energy than those containing limestone aggregates.
- Fracture energy increases with decrease in the air void content.
- Modified asphalt mixtures exhibit higher value of fracture energy compared to plain asphalt mixtures. This improvement in fracture energy is more pronounced at lower temperatures.
- Fracture energy increases as softer binder is used in asphalt mixture.
- Fracture energy decreases with increase in the loading rate.
- Fracture energy increases as the specimen thickness increases.
- Fracture energy decreases as larger aggregates are used in the mixture.
- Different specimens show different values of fracture energy.
- Fracture energy obtained from the mixed mod I/II test is higher than that obtained from the pure mode I test.

Critical J-integral is another fracture resistance indicator which is widely used by researchers to evaluate fracture behavior of asphalt mixtures. In order to calculate the critical value of J-integral, J_c, the specimens with at least two different notch lengths are required to be tested. Strain energy for each notch length is obtained by calculating the area under the load–displacement curve to the failure point. The strain energy versus the notch length is then plotted. The slope of the fitting line represents the critical J-integral, J_c value. It is notable that the strain energy reduces as the notch length increases.

Index

The manufacturer's authorised representative in the EU is Springer
Nature Customer Service Centre GmbH, Europaplatz 3, 69115 Heidelberg,
Germany. If you have any concerns regarding our products, please
contact ProductSafety@springernature.com

Printed and bound by CPI Group (UK) Ltd, Croydon, CR0 4YY

28/11/2025

02007709-0007